```
TJ           Oxley, P. L. B.
1185         (Peter Louis
.O82         Brennan), 1930-
1989
             The mechanics of
             machining.

$92.95
```

DATE			

BUSINESS/SCIENCE/TECHNOLOGY
DIVISION

© THE BAKER & TAYLOR CO.

THE MECHANICS OF MACHINING:
An Analytical Approach to Assessing Machinability

ELLIS HORWOOD SERIES IN MECHANICAL ENGINEERING

Series Editor: J. M. ALEXANDER, formerly Stocker Visiting Professor of Engineering and Technology, Ohio University, Athens, USA, and Professor of Applied Mechanics, Imperial College of Science and Technology, University of London

The series has two objectives: of satisfying the requirements of postgraduate and mid-career engineers, and of providing clear and modern texts for more basic undergraduate topics. It is also the intention to include English translations of outstanding texts from other languages, introducing works of international merit. Ideas for enlarging the series are always welcomed.

Alexander, J.M.	Strength of Materials: Vol. 1: Fundamentals; Vol. 2: Applications
Alexander, J.M., Brewer, R.C. & Rowe, G.	Manufacturing Technology Volume 1: Engineering Materials
Alexander, J.M., Brewer, R.C. & Rowe, G.	Manufacturing Technology Volume 2: Engineering Processes
Atkins A.G. & Mai, Y.W.	Elastic and Plastic Fracture
Beards, C.F.	Vibration Analysis and Control System Dynamics
Beards, C.F.	Structural Vibration Analysis
Beards, C.F.	Noise Control
Beards, C.F.	Vibrations and Control Systems
Besant, C.B. & Lui, C.W.K.	Computer-aided Design and Manufacture, 3rd Edition
Borkowski, J. and Szymanski, A.	Technology of Abrasives and Abrasive Tools
Borkowski, J. and Szymanski, A.	Uses of Abrasives and Abrasive Tools
Brook, R. and Howard, I.C.	Introductory Fracture Mechanics
Cameron, A.	Basic Lubrication Theory, 3rd Edition
Collar, A.R. & Simpson, A.	Matrices and Engineering Dynamics
Cookson, R.A. & El-Zafrany, A.	Finite Element Techniques for Engineering Analysis
Cookson, R.A. & El-Zafrany, A.	Techniques of the Boundary Element Method
Ding, Q.L. & Davies, B.J.	Surface Engineering Geometry for Computer-aided Design and Manufacture
Edmunds, H.G.	Mechanical Foundations of Engineering Science
Fenner, D.N.	Engineering Stress Analysis
Fenner, R.T.	Engineering Elasticity
Ford, Sir Hugh, FRS, & Alexander, J.M.	Advanced Mechanics of Materials, 2nd Edition
Gallagher, C.C. & Knight, W.A.	Group Technology Production Methods in Manufacture
Gohar, R.	Elastohydrodynamics
Gosman, B.E., Launder, A.D. & Reece, G.	Computer-aided Engineering: Heat Transfer and Fluid Flow
Gunasekera, J.S.	CAD/CAM of Dies
Haddad, S.D. & Watson, N.	Principles and Performance in Diesel Engineering
Haddad, S.D. & Watson, N.	Design and Applications in Diesel Engineering
Haddad, S.D.	Advanced Diesel Engineering and Operation
Hunt, S.E.	Nuclear Physics for Engineers and Scientists
Irons, B.M. & Ahmad, S.	Techniques of Finite Elements
Irons, B.M. & Shrive, N.G.	Finite Element Primer
Johnson, W. & Mellor, P.B.	Engineering Plasticity
Juhasz, A.Z. and Opoczky, L.	Mechanical Activation of Silicates by Grinding
Kleiber, M.	Incremental Finite Element Modelling in Non-linear Solid Mechanics
Kleiber, M. & Breitkopf, P.	Finite Element Methods in Structural Engineering: Turbo Pascal Programs for Microcomputers
Leech, D.J. & Turner, B.T.	Engineering Design for Profit
Lewins, J.D.	Engineering Thermodynamics
Malkin, S.	Materials Grinding: Theory and Applications
Maltbaek, J.C.	Dynamics in Engineering
McCloy, D. & Martin, H.R.	Control of Fluid Power: Analysis and Design, 2nd (Revised) Edition
Osyczka, A.	Multicriterion Optimisation in Engineering
Oxley, P.L.B.	The Mechanics of Machining
Piszcek, K. and Niziol, J.	Random Vibration of Mechanical Systems
Polanski, S.	Bulk Containers: Design and Engineering of Surfaces and Shapes
Prentis, J.M.	Dynamics of Mechanical Systems, 2nd Edition
Renton, J.D.	Applied Elasticity
Richards, T.H.	Energy Methods in Vibration Analysis
Ross, C.T.F.	Computational Methods in Structural and Continuum Mechanics
Ross, C.T.F.	Finite Element Programs for Axisymmetric Problems in Engineering
Ross, C.T.F.	Finite Element Methods in Structural Mechanics
Ross, C.T.F.	Applied Stress Analysis
Ross, C.T.F.	Advanced Applied Stress Analysis
Roy, D. N.	Applied Fluid Mechanics
Roznowski, T.	Moving Heat Sources in Thermoelasticity
Sawczuk, A.	Mechanics and Plasticity of Structures
Sherwin, K.	Engineering Design for Performance
Stupnicki, J.	Stress Measurement by Photoelastic Coating
Szczepinski, W. & Szlagowski, J.	Plastic Design of Complex Shape Structured Elements
Thring, M.W.	Robots and Telechirs
Walshaw, A.C.	Mechanical Vibrations with Applications
Williams, J.G.	Fracture Mechanics of Polymers
Williams, J.G.	Stress Analysis of Polymers 2nd (Revised) Edition

THE MECHANICS OF MACHINING:
An Analytical Approach to Assessing Machinability

P. L. B. OXLEY
Sir James Kirby Professor of Production Engineering
University of New South Wales
Kensington, Australia

ELLIS HORWOOD LIMITED
Publishers · Chichester

Halsted Press: a division of
JOHN WILEY & SONS
New York · Chichester · Brisbane · Toronto

First published in 1989 by
ELLIS HORWOOD LIMITED
Market Cross House, Cooper Street,
Chichester, West Sussex, PO19 1EB, England
The publisher's colophon is reproduced from James Gillison's drawing of the ancient Market Cross, Chichester.

Distributors:
Australia and New Zealand:
JACARANDA WILEY LIMITED
GPO Box 859, Brisbane, Queensland 4001, Australia

Canada:
JOHN WILEY & SONS CANADA LIMITED
22 Worcester Road, Rexdale, Ontario, Canada

Europe and Africa:
JOHN WILEY & SONS LIMITED
Baffins Lane, Chichester, West Sussex, England

North and South America and the rest of the world:
Halsted Press: a division of
JOHN WILEY & SONS
605 Third Avenue, New York, NY 10158, USA

South-East Asia
JOHN WILEY & SONS (SEA) PTE LIMITED
37 Jalan Pemimpin # 05–04
Block B, Union Industrial Building, Singapore 2057

Indian Subcontinent
WILEY EASTERN LIMITED
4835/24 Ansari Road
Daryaganj, New Delhi 110002, India

© **1989 P.L.B. Oxley/Ellis Horwood Limited**

British Library Cataloguing in Publication Data
Oxley, P. L. B. (Peter Louis Brennan), *1930–*
The mechanics of machining.
1. Metals. Machining. Mechanics
I. Title
671.3′5′

Library of Congress CIP data available

ISBN 0–7458–0007–6 (Ellis Horwood Limited)
ISBN 0–470–21382–5 (Halsted Press)

Typeset in Times by Ellis Horwood Limited
Printed in Great Britain by Hartnolls, Bodmin

COPYRIGHT NOTICE
All Rights Reserved. No part of this publication may be reproduced, stored in a retrieval system, or transmitted, in any form or by any means, electronic, mechanical, photocopying, recording or otherwise, without the permission of Ellis Horwood Limited, Market Cross House, Cooper Street, Chichester, West Sussex, England.

Table of contents

Preface 9

Symbols 13

1 Introduction
  1.1 General 17
  1.2 Reasons for studying mechanics of machining 17
  1.3 Orthogonal machining process 18
  1.4 Basic chip formation processes 19
  1.5 Basic problem requiring solution 21
  1.6 Historical 22

2 Shear plane and related solutions
  2.1 Shear plane model 23
  2.2 Shear plane solutions 26
  2.3 Curled chip solutions 32
  2.4 Restricted tool–chip contact solutions 34
  2.5 Limitations of perfectly plastic solutions 35

3 Slipline field analyses of experimental flow fields
  3.1 Introduction 37
  3.2 Experimental flow fields 38
  3.3 General observations from experimental flow fields 40
  3.4 Slipline field analyses 42
  3.5 Computer-aided methods 49

4 Parallel-sided shear zone theory
  4.1 Chip formation model 50
  4.2 Theory 51
  4.3 Effect of strain-hardening on shear angle 54
  4.4 Strain-rate effects 55
  4.5 Predicting the influence of cutting speed on shear angle and cutting forces 61

Table of contents

5 Experimental investigation of the influence of speed and scale on the strain-rate in the chip formation zone
- 5.1 Introduction ... 65
- 5.2 Experiment ... 66
- 5.3 Analysis of experimental flow fields ... 67
- 5.4 Results and discussion ... 70

6 Work material properties: the influence of strain-rate and temperature
- 6.1 Flow stress data determined from machining test results ... 74
- 6.2 Temperature calculation methods ... 78
- 6.3 Thermal properties ... 86
- 6.4 Interpretation of flow stress data in terms of strain-rate and temperature ... 87
- 6.5 Flow stress data determined from high speed compression test results ... 94

7 Predictive machining theory based on a chip formation model derived from analyses of experimental flow fields
- 7.1 Basic theory and calculation procedure ... 97
- 7.2 Determination of strain-rate in chip formation zone as part of solution ... 105
- 7.3 Nature of tool–chip interface friction and modelling it as a minimum work process ... 109
- 7.4 Comparison of predicted and experimental chip geometry and cutting force results ... 114

8 Predicting cutting forces for oblique machining conditions
- 8.1 Introduction ... 136
- 8.2 Determining the chip flow direction ... 137
- 8.3 Cutting force prediction methods ... 140
- 8.4 Comparison of predicted and experimental cutting force results ... 143

9 Allowing for end cutting edge effects in predicting cutting forces in bar turning with oblique machining conditions
- 9.1 Introduction ... 147
- 9.2 Experiment ... 147
- 9.3 Predicted results neglecting end cutting edge effects ... 148
- 9.4 Description of an equivalent cutting edge ... 151
- 9.5 Predicted results using equivalent cutting edge ... 158

10 Influence of tool nose radius on chip flow direction and cutting forces in bar turning
- 10.1 Introduction ... 160
- 10.2 Prediction of chip flow direction ... 160
- 10.3 Experimental investigation of chip flow direction and comparison with predicted results ... 176

    10.4 Prediction of cutting forces and comparison with experimental results 177

11 Assessing machinability factors in terms of machining theory
    11.1 Introduction 183
    11.2 Cutting forces and power 183
    11.3 Tool wear and tool life 184
    11.4 Cutting edge deformation 191
    11.5 Surface finish 197
    11.6 Further machinability considerations 202
    11.7 Concluding remarks 205

Appendix A1 Elements of plane strain slipline field theory 208
    A1.1 Plane strain equations referred to Cartesian coordinates 208
    A1.2 Plane strain equations referred to sliplines 211
    A1.3 Outline of method for obtaining perfectly plastic solutions 214

Appendix A2 Relating plane strain and uniaxial conditions and determining the shear flow stress distribution for experimental flow fields . . . 220

Appendix A3 Representation of σ_1 and n as functions of velocity-modified temperature and carbon content for a range of plain carbon steels 223

References 228

Index 239

To Beryl

Preface

Traditionally the selection of cutting tools and cutting conditions (speed and size of cut) in machining processes such as turning and milling has been left to the machine tool operator who would have acquired the necessary skills from practical experience. The operator's responsibility would also include the monitoring of the process to make such decisions as to when the tool needs replacing or regrinding, or when the cutting speed should be changed to allow for unexpected variations in work material hardness, etc. Although even today this situation would still apply in many machine shops the position is changing rapidly. Increasingly the type of tool material, tool geometry and cutting conditions are being selected in the planning or some such department on the basis of experimentally obtained machining data provided by research organisations and by the exchange of know-how within manufacturing industry itself. To this end many countries are at present setting up machinability data banks to collect and collate information on such factors as tool performance and cutting forces and hence power requirements for wide ranges of work and tool materials. In this way it is hoped to operate machining processes near to optimum conditions from the viewpoint of cost and productivity. The increasing use of numerically controlled machine tools and the replacement of manual by computer control makes the selection of cutting conditions which will ensure that the cutting edge will remove material from the workpiece effectively for an acceptable length of time even more pressing. This emphasises the need for reliable machining data and lacking this it is clear that there will be a tendency to run machines at conservative cutting speeds, etc., well below the optimum values in order that the cutting tool does not fail prematurely and damage a component on which much work has already been done.

Unfortunately the obtaining of statistically reliable experimental machining data is extremely costly in terms of both time and material. The difficulty of this task is made even clearer when the wide range of work and tool materials in use is considered and when it is noted how even slight variations in the chemical composition or metallurgical state of nominally the same work material causes very large changes in tool performance, etc. Although the application of empirical relations such as the widely used Taylor tool life equation help in reducing the experimental work it is clear that much greater headway could be made if tool life and other machinability parameters could be related more fundamentally to work and tool material properties and cutting conditions. It is this objective which has led many researchers for the past 100 years or so to investigate the basic mechanics of the machining process. The author has been working in this field of research for over 30 years and this book is essentially an account of this work.

The main differences between the author's work and that of other investigators are in the emphasis placed on taking as much account as possible of plasticity theory in analysing the chip formation process by which material is removed from the workpiece and in introducing into the analysis the temperature and strain-rate dependent work material flow stress and thermal properties. In this way a machining theory has been developed which enables predictions to be made of cutting forces, temperatures, stresses, etc., from a knowledge of the work material properties and cutting conditions which are in good agreement with experimental values. Once the tool temperatures and stresses are known then it is possible to predict such factors as tool life and the cutting conditions which give plastic deformation of the cutting edge. Predictions can also be made of the cutting conditions which give a built-up edge and are therefore to be avoided when a good surface finish is required. Originally the theory was limited to the relatively simple two-dimensional case of orthogonal machining but has now been extended to the more general three-dimensional case of oblique machining. In both these cases cutting was initially assumed to occur only along a single, straight, cutting edge but recently account has been taken of cutting on more than one edge as in bar turning and of the influence of the tool nose radius. Preliminary attempts have also been made to apply the theory to more complicated processes such as milling and grinding.

After a brief introductory chapter a description is given of the highly idealised shear plane and related slipline field analyses of chip formation in which little account is taken of real work material properties. These are, however, a valuable starting point for the author's own work, the development of which is then traced out in more or less historical order. The practical value of the work is highlighted where appropriate and particularly in the last chapter which considers the application of the theory to making machinability assessments.

The approach used in the book is to apply continuum mechanics as far as possible to the analysis of the chip formation process. Neglect of the metallurgical aspects of the process does not mean that they are unimportant, they are of course vitally important, but simply reflects the author's own research interests. Fortunately other authors have looked at these aspects. Also no attempt has been made to review the vast amount of work on the mechanics of machining carried out by others and only that closely related to the author's own approach is referred to. The book should be particularly relevant to postgraduates and final year undergraduates specialising in manufacturing processes but should also be useful to those in manufacturing industry who are charged with the responsibility of making machining processes ever more efficient. After all, however sophisticated the machine tools and control systems used in machining operations the actual happenings right at the cutting edge are of paramount importance.

I am deeply indebted to the following people for their help with the work

Preface

described in this book: Professor Sir Derman Christopherson for suggesting the machining problem to me and starting me on my research career, and my joint supervisor Dr Bill Palmer particularly for his painstaking correction of my thesis and for giving me my first serious lesson in technical writing; my colleagues Dr Eleonora Kopalinsky and Professor Jack Taylor for valuable discussions and criticisms of the work in more recent years; my research students, whose names appear in the references throughout the book, without whose efforts far less would have been achieved; and Ron Fowle, Ron Montgomery and Osman Savci for their expert contribution to the experimental work. I wish to thank Catherine Faust for producing the excellent line drawings given in the figures and Xiaoping Li for his careful checking of the manuscript. Finally I wish to thank Diane Augee whose word processing and organisational expertise and ever-cheerful approach to problems have made the writing of this book far less arduous than it might otherwise have been.

Symbols

a, b	constants in tool life equation
$\hat{a}, \hat{b}, \hat{c}$ etc.	unit vectors
A, B	constants in tool life–tool temperature relation
A_r	cutting face plane
C, C'	strain-rate constants for chip formation zone
	constant in tool life equation
C_e	end cutting edge angle
C_s	side cutting edge angle
d	depth of cut
dA	area of undeformed section of chip element (nose radius tool)
$d\boldsymbol{F}$	friction force acting on chip element (nose radius tool)
dr	differential length of cutting edge for circular arc cutting edge
ds	differential length of cutting edge for cutting edge of arbitrary shape
E	elastic modulus
f	feed
	normalised interfacial film strength
F	frictional force at tool–chip interface
$\boldsymbol{F}$	resultant friction force at tool–chip interface (nose radius tool)
F_C	force in cutting direction
F_N	force normal to AB
F_R	force normal to F_C and F_T
F_S	shear force on AB
F_T	force normal to cutting direction and machined surface
F_X, F_Y	forces in X and Y directions (nose radius tool)
h	tool–chip contact length
i	inclination angle
k	shear flow stress
	constant in chip flow direction equation
k_{AB}, k_{CD}, k_{EF}	shear flow stresses on AB, CD and EF

Symbols

k_{chip} (k_{int})	shear flow stress in chip at tool–chip interface
k_0	shear flow stress at zero plastic strain
Δk	change in shear flow stress in parallel-sided shear zone
K	thermal conductivity
l	length of AB
lg	logarithm to base 10
ln	logarithm to base e
L	tool life
m	slope of linear plastic stress–strain relation
n	strain-hardening index
N	normal force at tool–chip interface
p	hydrostatic stress
p_A, p_B	hydrostatic stresses at A and B
p'_A, p'_C	hydrostatic stresses at points A and C used in deriving Hencky's first theorem
p_a	average hydrostatic stress on AB
P_1, P_2, P_3	three mutually perpendicular components of force for case when $C_s \neq 0°$
P_f, P_p, P_r, etc.	planes used in defining equivalent cutting edge
r	nose radius
R	resultant force
R_T	thermal number
s	slope of linear shear flow stress against normal stress relation
	slope of linear shear fracture strength against hydrostatic stress relation
s_1, s_2	distances measured along I and II sliplines
Δs_2	thickness of parallel-sided shear zone
S	specific heat
t	undeformed chip thickness for circular arc cutting edge
$t(s)$	undeformed chip thickness for cutting edge of arbitrary shape
t_1	undeformed chip thickness
t_2	chip thickness
T_{AB}	average temperature along AB
T_C	mean chip temperature
T_{flank}	tool flank temperature
T_{int}	average temperature along tool–chip interface
T_{mod}	velocity-modified temperature
T_{tool}	tool temperature
T_W	initial work temperature
ΔT	temperature rise
ΔT_C	average temperature rise in chip
ΔT_M	maximum temperature rise in chip
ΔT_{SZ}	temperature rise in chip formation zone
u, v	velocities along I and II sliplines
$\mathbf{u}$	friction force intensity
U	cutting velocity
v_x, v_y, v_z	Cartesian components of velocity

Symbols

v_p, v_q	velocities at points p and q
V	chip velocity
V_N	velocity normal to AB
V_S	velocity discontinuity
	shear velocity
ΔV_S	change in shear velocity across an element of thickness Δs
w	width of cut
x, y, z and X, Y, Z	Cartesian coordinates
α	rake angle
α_e	effective rake angle
α_n	normal rake angle
β	proportion of heat conducted into work
	clearance angle
	angle turned through by streamline
	slipline field angle
$\Delta\beta$	angle turned through by streamline in crossing element
γ	shear strain
	slipline field angle
$\gamma_{AB}, \gamma_{CD}, \gamma_{EF}$	shear strains at AB, CD and EF
γ'_{AB}	shear strain rise in reaching AB in case of cold worked material
γ_f	fracture shear strain at zero hydrostatic stress
γ_{SP}	shear strain occurring in crossing shear plane
γ_w	shear strain resulting from cold working
$\Delta\gamma$	shear strain occurring across element
$\dot{\gamma}$	shear strain-rate
$\dot{\gamma}_{xy}, \dot{\gamma}_{yz}, \dot{\gamma}_{zx}$	Cartesian components of shear strain-rate
$\dot{\gamma}_{AB}$	shear strain-rate at AB
$\dot{\gamma}_{int}$	shear strain-rate at tool–chip interface
$\dot{\gamma}_{max}$	maximum shear strain-rate
$\dot{\gamma}_{SZ}$	shear strain-rate in parallel-sided shear zone
δ	ratio of tool–chip interface plastic zone thickness to chip thickness
ε	uniaxial (effective) strain
$\bar{\varepsilon}$	effective strain
ε_{AB}	effective strain at AB
ε_{int}	effective strain at tool–chip interface
ε_w	effective strain resulting from cold working
$\dot{\varepsilon}$	uniaxial (effective) strain-rate
$\dot{\bar{\varepsilon}}$	effective strain-rate
$\dot{\varepsilon}_{AB}$	strain-rate at AB
$\dot{\varepsilon}_{int}$	strain-rate at tool–chip interface
$\dot{\varepsilon}_x, \dot{\varepsilon}_y, \dot{\varepsilon}_z$	Cartesian components of direct strain-rate
$\dot{\varepsilon}_1, \dot{\varepsilon}_2, \dot{\varepsilon}_3$	principal strain-rates
ζ	slipline field angle
η	chip formation zone temperature factor
	slipline field angle

Symbols

	centred fan angle giving size of built-up edge
	angle between shear plane direction and direction of maximum shear stress
	angle of chip flow in restricted tool–chip contact model
η_c	chip flow angle measured from normal to side cutting edge
θ	angle made by R with AB
	slipline field angle
	angle of sliplines determined by τ
$\theta_1, \theta_2, \theta_3$	limits of integration
λ	mean friction angle
μ	coefficient of friction
$\nu, \dot{\varepsilon}_0$	constants in velocity-modified temperature equation
ρ	density
σ	uniaxial (effective) flow stress
σ_1	value of σ at $\varepsilon = 1$
$\bar{\sigma}$	effective stress
σ_{AB}	flow stress at AB
σ_{int}	flow stress at tool–chip interface
σ_N, σ'_N	normal stress acting on tool–chip interface at B
σ_m	mean (hydrostatic) component of stress
$\sigma_x, \sigma_y, \sigma_z$	Cartesian components of direct stress
$\sigma'_x, \sigma'_y, \sigma'_z$	deviatoric stress components
$\sigma_1, \sigma_2, \sigma_3$	principal stresses
τ	frictional shear stress
$\tau_{xy}, \tau_{yz}, \tau_{zx}$	Cartesian components of shear stress
τ_{int}	resolved shear stress at tool–chip interface
τ_{max}	maximum shear stress
ϕ	shear angle
Φ	angle made by AB with U in built-up edge model
	angle made by CD with U in wave model of asperity deformation
ψ	angular rotation of I sliplines from x-axis
	tool–chip interface temperature factor
	slipline field angle
$\psi_A, \psi_B, \psi_C, \psi_D$	values of ψ at points A, B, C, D used in deriving Hencky's first theorem
Ω	chip flow angle measured from tool axis
$\star$	indicates angles and planes referred to equivalent cutting edge

1
Introduction

1.1 GENERAL

The machining processes considered in this book are of the type in which a component is produced by the removal of material from a workpiece by the action of the wedge-shaped cutting edge (or edges) of the tool used. The removed material is normally called swarf or chips with the latter name preferred in machining research. Typical machining processes of this kind include turning, milling, drilling, shaping, broaching and grinding. Metal is the most widely machined of all materials and it is quite common for machining processes to be called metal cutting processes. Irrespective of which title is used the conditions such as speed under which a process is carried out and the forces involved are usually referred to as cutting conditions and cutting forces. This tradition is continued in the present work. Processes such as electrodischarge, electrochemical, electron beam and ultrasonic machining which in the current sense are non-chip-forming processes are not considered. Nor, for the same reason, are the sheet metal cutting processes of guillotining, etc.

Attention is limited to metal work materials. Indeed machining theory has only been developed in any detail for metals. The methods of analysis used are essentially those of continuum mechanics. Metallurgical aspects of machining are largely ignored: this does not mean that they are unimportant but rather that they lie outside the scope of the present treatment. Trent (1977) has given an excellent account of this side of the problem.

1.2 REASONS FOR STUDYING MECHANICS OF MACHINING

In mechanics of machining research it is the basic chip formation process by which material is removed from the workpiece which is studied. The purpose of this research is to provide a theory of machining which relates cutting forces, tool stresses and temperatures, etc., to the cutting conditions (cutting speed, size of cut and cutting edge geometry) and work and tool material properties.† It should then be possible to determine such practically important factors as the cutting power from the forces, and the effective life of the cutting tool from the tool temperatures and stresses. At present it is usual to use empirical relations for this purpose. In this connection Taylor (1907), Koenigsberger (1964), Kronenberg (1966) and others including organisations such as Metcut (1980) have made experimental measurements of tool life, cutting forces, etc., and have presented their results in a form suitable for industrial application. The collection of data of this kind is now being undertaken in many countries in order to set up machinability data banks — see, for example, the CIRP report (1976) on this topic. Unfortunately this process is extremely time consuming and costly. This becomes especially important when the constant introduction of new work and tool materials is considered and when it is realised how even small deviations from the nominal composition of a work material can cause large changes in its machining characteristics. To reduce this work it is clear that more fundamental relationships than purely empirical ones are needed. The aim of research into the mechanics of machining is to provide these.

1.3 ORTHOGONAL MACHINING PROCESS

In both experimental and analytical investigations of chip formation it has been usual to consider the relatively simple case of orthogonal machining. In this process as shown in Fig. 1.1 a tool with a plane cutting face and a single, straight cutting edge,

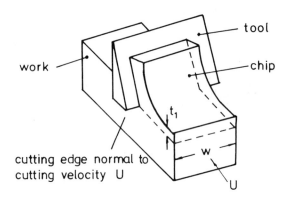

Fig. 1.1 — Orthogonal machining process.

which is set normal to the cutting velocity U (velocity of work relative to tool), removes a layer of work material of uniform thickness t_1 and width w. The geometry of the cutting edge can in this case be defined by its width, which must be greater than

---

† Any comprehensive theory should also take account of the influence of cutting fluids on the process but to date this has only been allowed for in a relatively crude way. Virtually all of the work described in this book is therefore for dry conditions although some indication is given in the final chapter of how allowance might be made in the theory for the influence of cutting fluids.

the width of cut, and by the two angles α and β (Fig. 1.2). The angle α between the tool cutting face and the normal to the cutting velocity U is termed the rake angle; it is measured positive as shown in Fig. 1.2(a) and negative as shown in Fig. 1.2(b). Experiments show that the rake angle has a profound effect on the chip formation process and hence on cutting forces, etc. The angle β between the clearance face of the tool and the work surface is termed the clearance angle. It is generally considered to be of minor importance in the mechanics of chip formation although as shown by Taylor (1955) it can influence the rate of clearance face wear. In practice the value of the clearance angle is determined by cutting edge strength considerations and the need for the tool to clear the machined surface.

Although many practical machining processes approximate quite closely to orthogonal machining conditions it is clear that processes could often be represented more accurately by a model in which the cutting edge is not normal to the cutting velocity and cutting is on more than one edge. It will be seen in later chapters how this can be achieved using the orthogonal model as a basis. Initially, however, attention will be concentrated on the orthogonal machining process.

1.4 BASIC CHIP FORMATION PROCESSES

Photomicrographs of chip sections have been used extensively to observe the different types of material removal process which can take place in machining. In this method a cut is taken under given conditions and stopped as quickly as possible so that the observed deformation is near to that while actually machining — see Hastings (1967) for a description of 'quick-stop' devices used for this purpose. The workpiece is then removed and a section, in a plane parallel to the direction of cutting and perpendicular to the machined surface, polished and etched; a magnified picture is then taken of the zone of deformation between work and chip. Using this approach Rosenhain and Sturney (1925) appear to have been among the first to attempt to classify the types of chip which were produced. They suggested three basic types of chip which they termed 'tear', 'shear' and 'flow' types.

The work of Rosenhain and Sturney was superseded by a later classification due to Ernst (1938) who used high speed motion pictures, in addition to photomicrographs, to distinguish three types of chip, namely discontinuous, continuous and continuous with built-up edge. Sketches of these are given in Fig. 1.3. They can be described as follows. With the discontinuous chip (Fig. 1.3(a)) fracture is observed to occur ahead of the tool and the chip has the appearance of being made up of many segments. The resulting machined surface is rough and irregular. This type of chip usually occurs when machining brittle materials, although under certain conditions, for example slow cutting speeds and high negative rake angles, it can also occur even when machining relatively ductile materials. With the continuous chip (Fig. 1.3(b)) the chip is formed by plastic deformation with no fracture, at least on the macro scale. The machined surface in this case is smooth. The continuous chip is usually obtained when machining ductile materials at reasonably high cutting speeds. With the continuous chip with built-up edge (Fig. 1.3(c)) photomicrographs indicate that a cap of highly deformed material is formed on the tool face adjacent to the cutting edge. This is the built-up edge which can sometimes be observed 'welded' to the cutting edge after a machining operation. Using motion pictures Ernst showed that

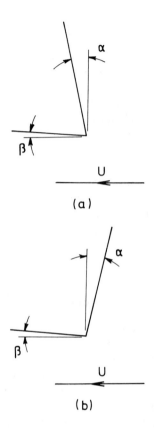

Fig. 1.2 — Cutting edge geometry: (a) positive rake angle; (b) negative rake angle.

built-up edge formation is cyclic in nature with the edge building up to a certain size before becoming unstable and then breaking off. Part of the built-up edge then goes to form the underside of the chip and part to form the newly machined surface. The resulting surface finish in this case is poor. Built-up edge tends to occur at intermediate cutting speeds when machining ductile material but it is difficult to specify simple rules for its occurrence. Ernst suggested that built-up edge was caused by the high value of friction between chip and tool resulting in shear failure on the surface of the chip as it rubbed the tool face. He concluded that the value of tool–chip interface friction was of considerable importance in determining the type of chip produced. Ernst's classification of chip types is of course not all encompassing. It does not, for example, include the catastrophic shear type chip obtained when machining materials such as titanium and stainless steel and even more commonplace materials such as plain carbon steels under extreme conditions. Nor does it subclassify the various forms of built-up edge which can occur as shown by Heginbotham and Gogia (1961). It is nevertheless a most useful classification.

It can be seen that only when machining with a continuous chip (Fig. 1.3(b)) can the process approximate to a steady-state process. Not surprisingly therefore it is this process which is assumed to apply in most machining analyses. When the chip is

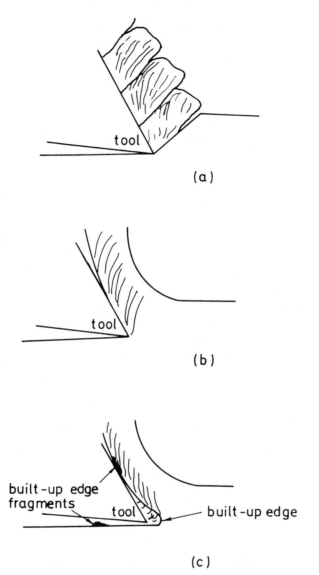

Fig. 1.3 — Sketches of different chip types (Ernst 1938): (a) discontinuous; (b) continuous; (c) continuous with built-up edge.

discontinuous or a built-up edge occurs the processes are non-steady-state and would appear prohibitively difficult to analyse. However, as will be seen, machining theory based on a steady-state model (continuous chip) can be used to predict when this model no longer applies and when, for example, a built-up edge is likely to occur.

1.5 BASIC PROBLEM REQUIRING SOLUTION

If attention is limited to machining with a continuous chip then the process is essentially one of plastic deformation and the appropriate theory for its solution is plasticity theory.† For problems of large plastic strains such as machining perhaps the most powerful theory available for obtaining solutions is plane strain slipline field theory. Slipline field methods are widely used in this book and a description of slipline field theory is given in Appendix A1.

To achieve true plane strain conditions in orthogonal machining with no flow in the direction of the cutting edge would require a specially designed experiment with side plates used to prevent the flow. However, machining experiments show that plane strain conditions are approximately met if the width of cut w (Fig. 1.1) is large compared with its thickness t_1. In particular if $w \geqslant 10 t_1$ then plane strain conditions apply across most of the width and side flow is restricted to narrow regions at the edges. The reason for this is that the 'rigid' work material adjacent to the deforming region limits side spread. In the orthogonal machining analyses to be considered the dimensions of the cut are presumed to be such as to make the assumption of plane strain conditions reasonable.

The problem to be solved can be stated as follows. Given the cutting conditions and the appropriate work and tool material properties, determine the geometry of the chip and of the zones of plastic deformation together with the associated stresses, strains, strain-rates and temperatures. No complete solution to the problem as formulated has been obtained even for the idealised rigid–perfectly plastic (non-hardening) work material normally assumed in slipline field theory. Indeed in this context Hill (1954) has questioned the existence of a unique solution for a given set of conditions. It is doubtful if a complete solution to the problem can be found even with the powerful numerical methods now available, although Usui and Shirakashi (1982) and Childs and Maekawa (1987) have made commendable attempts to do this using finite element methods. In any case it is possible that the computer time involved would be excessive and that numerical methods could well be uneconomic even when compared with the empirical approach mentioned earlier. To date the usual approach to the problem has been to propose a chip formation model based on experimental observations and then to develop an approximate machining theory from this. The best known model of this kind is the shear plane model of chip formation and this and the associated shear plane solutions are considered first.

1.6 HISTORICAL

Machining research has been carried out for well over 100 years and there is a vast literature on the topic. Reviews which taken together cover work from the early period up to the 1950s have been given by Finnie (1956), Zorev (1966) and Shaw (1968).

---

† It should be noted that inertial stresses only become significant in the chip formation process at cutting speeds greatly in excess of the practical range and are therefore normally neglected in machining analyses.

2

Shear plane and related solutions

2.1 SHEAR PLANE MODEL

The shear plane model of chip formation (Fig. 2.1) is based on the experimental

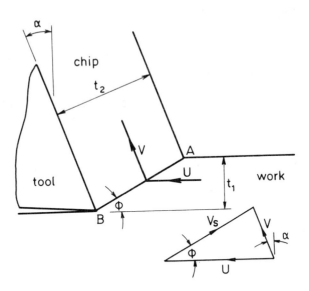

Fig. 2.1 — Shear plane model.

observation made by Ernst (1938) and others that the continuous chip is formed by plastic deformation in a narrow zone that runs from the tool cutting edge to the work–chip free surface. This is represented by the shear plane AB in Fig. 2.1 across which the work velocity U (the tool is assumed stationary) is instantaneously changed to the chip velocity V. This requires a discontinuity (jump) in the tangential component of velocity across AB equal to V_S as shown by the velocity diagram in Fig.

2.1. The model as described is only valid for an idealized rigid–perfectly plastic (non-hardening) work material. For such a material the elastic strain is disregarded and during deformation the volume of an element remains constant. Conservation of mass therefore requires that the normal component of velocity is continuous across AB as the velocity diagram shows it to be, that is the components of U and V normal to AB are equal. No such restriction is placed by this consideration on the tangential component. If, however, the material hardens during deformation then as will be seen later the discontinuity in tangential velocity is no longer permissible. In this case AB must open up to form a finite shear zone through which elements of material flow along smooth streamlines.

The shear plane AB as a plane of tangential velocity discontinuity is a direction of maximum shear strain rate ($\dot{\gamma}_{AB} \to \infty$ in this case) and hence from isotropic plasticity theory can also be assumed to be a direction of maximum shear stress. In terms of plane strain slipline field theory for a rigid–perfectly plastic material (see Appendix A1) AB is a slipline. The stress components acting at a point on a slipline are the mean compressive (hydrostatic) stress p which acts normal to the slipline and the shear flow stress k which acts parallel to the slipline. For a perfectly plastic material k remains constant during deformation and the stress equilibrium equations referred to sliplines (the Hencky equations) can be written in the form

$$p + 2k\psi = \text{constant along a I line} \\ p - 2k\psi = \text{constant along a II line} \tag{2.1}$$

where ψ is the anticlockwise angular rotation of the I lines from a fixed reference axis with the I lines taken as those on which the shear stress exerts a clockwise couple. From equations (2.1) the variation of p along a slipline is directly related to the angle turned through by the slipline. Hence if the shear plane AB is taken to be straight, as it is in most shear plane theories, then p is constant along AB and the resultant force transmitted by AB (R in Fig. 2.2) passes through its mid-point. If the tool is assumed

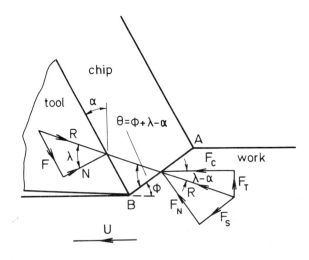

Fig. 2.2 — Forces associated with shear plane model.

Shear plane model

to be perfectly sharp, then the tool–chip interface (along which the chip and tool are in contact) must also transmit the same resultant force R.

It is convenient to split R into various sets of components as shown in Fig. 2.2 from which the following relations can be obtained

$$F_C = R \cos(\lambda - \alpha)$$
$$F_T = R \sin(\lambda - \alpha)$$
$$F = R \sin \lambda \qquad (2.2)$$
$$N = R \cos \lambda$$

$$R = \frac{F_S}{\cos \theta} = \frac{k_{AB} \, t_1 \, w}{\sin \phi \, \cos \theta}$$

where F_C and F_T are the forces in the cutting direction (direction of U in Fig. 2.2) and normal to this direction, F and N are the frictional and normal forces at the tool–chip interface, F_S is the shear force along AB, α is the tool rake angle, λ is the mean angle of friction used to describe the frictional condition at the tool–chip interface, ϕ is the angle made by AB with the cutting direction (the so-called shear angle), θ is the angle made by the resultant R with AB, k_{AB} is the shear flow stress along AB, t_1 (Fig. 2.1) is the undeformed chip thickness and w is the width of cut measured along the cutting edge.

Some useful relationships between velocities can be obtained from the velocity diagram in Fig. 2.1, namely

$$V = \frac{U \sin \phi}{\cos(\phi - \alpha)}$$
$$V_S = \frac{U \cos \alpha}{\cos(\phi - \alpha)} \qquad (2.3)$$
$$V_N = U \sin \phi$$

where V_N is the velocity normal to AB.

It can easily be shown that the shear strain occurring in crossing a tangential velocity discontinuity is given by the magnitude of the discontinuity divided by the magnitude of the component of velocity normal to the discontinuity. Hence the shear strain occurring as material crosses the shear plane AB (Fig. 2.1) is given by $\gamma_{SP} = V_S/V_N$ which on substituting for V_S and V_N from equations (2.3) gives

$$\gamma_{SP} = \frac{\cos \alpha}{\sin \phi \, \cos(\phi - \alpha)} \qquad (2.4)$$

2.2 SHEAR PLANE SOLUTIONS

For a given rake angle α and undeformed chip thickness t_1 it can be seen that the geometry of Fig. 2.1 is not completely defined unless the shear angle ϕ (or the chip thickness t_2) is known and hence no estimate can be made of cutting forces, etc. Many equations have been proposed for predicting ϕ and some of these are now described.

Ernst and Merchant (1941) expressed the shear stress along the shear plane in terms of R, w, t_1, α, λ and ϕ and then, making a number of simplifying assumptions, selected ϕ to make AB a direction of maximum shear stress which gave

$$\phi = \frac{\pi}{4} + \frac{\alpha}{2} - \frac{\lambda}{2} \tag{2.5}$$

Merchant (1945) later showed that this equation could also be derived by expressing the cutting force F_C in terms of the shear flow stress along AB (assumed constant), w, t_1, α, λ and ϕ and then selecting ϕ to make the expenditure of work in cutting a minimum. He then repeated the analysis but with the shear flow stress on AB allowed to increase linearly with increase in the normal stress on AB. This gave

$$2\phi = \cot^{-1} s + \alpha - \lambda \tag{2.6}$$

where s is the slope of the shear flow stress against normal stress relation. In the light of more recent work (Oxley 1963a) it can be seen that Merchant was in effect obtaining upper bound solutions to the problem. Strictly speaking the upper bound method is not applicable to problems where the boundary forms part of the solution as in machining. It is unfortunate that the method has been widely misapplied in this sense.

Lee and Shaffer (1951) proposed a shear angle solution based on the slipline field model of chip formation given in Fig. 2.3. In this all of the deformation is again taken to occur on the shear plane AB with the same velocity diagram (hodograph) and force diagram as in Figs. 2.1 and 2.2 applying. However, the plastic field is assumed to extend above AB with sliplines parallel to and normal to AB as shown. From equations (2.1) and (A1.16) the region ABC in which ψ is constant is therefore one of constant stress and moves as a rigid body although in a plastic stress state. If the additional assumption is made that BC represents the full contact length between chip and tool then AC is a stress-free surface. From equilibrium considerations (see Appendix A1) the sliplines must meet AC at an angle $\pi/4$ and p must be equal to k and compressive along this surface. From equations (2.1) it follows that $p = k$ along AB and hence that the angle made by the resultant cutting force R with AB (θ in Fig. 2.2) is $\pi/4$. From the geometry of Fig. 2.2

$$\theta = \phi + \lambda - \alpha \tag{2.7}$$

and therefore for the Lee and Shaffer model

Sec. 2.2] **Shear plane solutions** 27

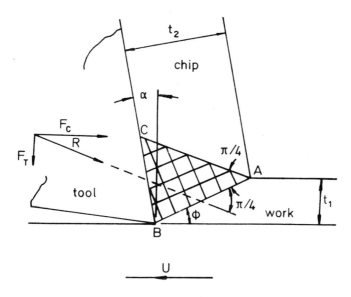

Fig. 2.3 — Lee and Shaffer (1951) model.

$$\phi = \frac{\pi}{4} + \alpha - \lambda \qquad (2.8)$$

Lee and Shaffer regarded this as their main solution but they showed that it was inapplicable under some conditions. In the first place when the rake angle is greater than the friction angle ($\alpha > \lambda$) the transmission of the tool force into the work does not give critical stress conditions in the chip, that is the material above AB is not in a plastic stress state. Their slipline field then degenerates into the single shear plane AB and they concluded that in such cases equation (2.5) would apply because it gave the least cutting force. Secondly an examination of equation (2.8) showed that for high values of λ and small or negative values of α a negative value of ϕ is implied. A negative value of ϕ is not physically possible and this led Lee and Shaffer to the conclusion that under such conditions a small permanent built-up edge exists. They assumed that this would be stable in character and could be likened to a cap of dead metal which was formed early in cutting and then remained constant in shape and size. The slipline field and hodograph suggested by Lee and Shaffer for this case are given in Fig 2.4. In the slipline field, BDE represents the built-up edge, the size of which, for a given value of undeformed chip thickness t_1, is defined by the centred fan angle η. The hodograph shows that AB is again a plane of tangential velocity discontinuity with the magnitude of the discontinuity depending on the slope of BD (which is both a slipline and a streamline) at B. There is further deformation within the fan region ABD as shown by the hodograph in which the circular arc is the orthogonal image of BD and the other circular arc sliplines in the fan. If the shear stress along CD is equal to k then BD meets CD tangentially and there is no discontinuity in tangential velocity across AD. If, however, this stress is less than k

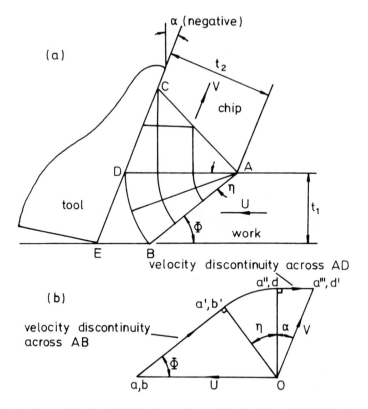

Fig. 2.4 — Lee and Shaffer (1951) built-up edge model: (a) slipline field; (b) hodograph. The velocities at points represented by capital letters in the slipline field are measured in the hodograph from the pole O to the corresponding lower case letters. Primes indicate where the velocity at a point has more than one value. Thus Ob and Ob' represent the velocities on the work side and plastic zone side respectively of B. In the case of point A the velocity changes instantaneously from U to V in three steps as indicated by the hodograph.

then there is a discontinuity as indicated in Fig. 2.4. If it is again assumed that no force is transmitted above the point C then by the same argument as before the resultant force across CD must make an angle $\pi/4$ with AD. It then follows from the geometry of Fig. 2.4 that

$$\Phi = \frac{\pi}{4} + \eta + \alpha - \lambda \tag{2.9}$$

where λ is the mean friction angle along CD and Φ is measured as shown. Lee and Shaffer gave a graphical method of estimating η in which it was required to know the mean angle of friction along BE in addition to α and λ. However, they did not attempt independent measurements of this factor.

To check how accurately shear angle equations predict ϕ it is necessary to determine α and λ experimentally. This can be accomplished by making orthogonal machining experiments in which the chip thickness t_2 (Fig. 2.1) and the cutting forces F_C and F_T (Fig. 2.2) are measured. It is usual to carry out such experiments on a lathe with a tubular workpiece reduced in length by a straight-edged cutting tool with the cutting edge set normal to the cutting velocity and feed direction which is parallel to the axis of the tube. To ensure that the conditions approximate closely to plane strain the width of cut w (tube wall thickness) should be at least ten times the undeformed chip thickness t_1 (axial feed) as mentioned earlier. Also, to ensure that the cutting velocity U is very nearly constant across the width of cut, w should be much smaller than the tube diameter. To reduce the labour involved in preparing workpieces a bar is sometimes used instead of a tube. Fortunately experiments show that for the kind of conditions described ($w \geq 10t_1$) the results obtained using a bar differ little from those obtained using a tube (Hastings *et al.* 1980). The forces F_C and F_T are measured using a dynamometer. (Shaw (1984) has given a comprehensive description of the design, etc., of cutting force dynamometers.) The chip thickness t_2 is measured from chips collected during a test. This measurement can be made directly using a ball-ended micrometer or alternatively it can be determined by weighing a given length of chip and knowing the density of the work material. This latter method is often preferred as it gives an average value which allows for the irregularities which are always observed in the chip outer surface. Knowing F_C and F_T, λ is found from the relation

$$\tan(\lambda - \alpha) = \frac{F_T}{F_C} \qquad (2.10)$$

which is obtained from equations (2.2). The relation for calculating ϕ from experimental values of t_2 is obtained from the geometry of Fig. 2.1. For plane strain, volume constancy requires that

$$Ut_1 = Vt_2$$

or

$$\frac{t_1}{t_2} = \frac{V}{U}$$

and substituting for V from equations (2.3) gives

$$\frac{t_1}{t_2} = \frac{\sin \phi}{\cos(\phi - \alpha)}$$

hence, expanding $\cos(\phi - \alpha)$ and rearranging,

$$\tan \phi = \frac{(t_1/t_2) \cos \alpha}{1 - (t_1/t_2) \sin \alpha} \qquad (2.11)$$

Experimental results obtained by Kobayashi and Thomsen (1959) for a range of work materials are given in Fig. 2.5. These are plotted in a form that has become

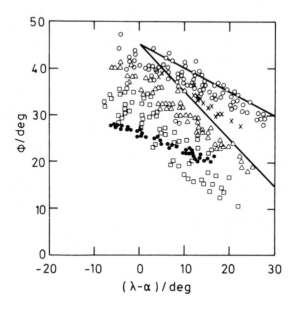

Fig. 2.5 — Experimental shear angle results (Kobayashi and Thomsen 1959): top line represents equation (2.5), bottom line represents equation (2.8).

Work material	Symbol
2024-T4 aluminium alloy	○
6061-T6 aluminium alloy	×
SAE 1112 steel (as received)	△
SAE 1112 steel (annealed)	□
Alpha brass	●

traditional, namely ϕ against $\lambda - \alpha$. On this basis equations (2.5) and (2.8) can be represented by straight lines as shown in Fig. 2.5. In both cases the intercept at $\lambda - \alpha = 0$ is $\pi/4$ and equation (2.5) has a slope of $-1/2$ while equation (2.8) has a slope of -1. If the experimental results for a particular material are considered (Fig. 2.5) it can be seen that generally speaking an increase in $\lambda - \alpha$ leads to a decrease in ϕ which is the result predicted by equations (2.5) and (2.8). The quantitative agreement is, however, very poor. In particular equations (2.5) and (2.8) each predict a unique value of ϕ for a given value of $\lambda - \alpha$ while the experimental results show that ϕ can vary by over 20°. Better agreement between predicted and experimental results can be obtained if equation (2.6) is used with s allowed to vary. Unfortunately the required variation is physically unacceptable. Although Φ in the Lee and Shaffer built-up edge model (Fig. 2.4) is not the same angle as ϕ in the shear plane model

(Fig. 2.1) and calculated from equation (2.11) it can be seen that these angles are related and that an increase in Φ by reducing t_2 would give an increase in the apparent value of ϕ. Equation (2.9) might therefore by suitable choice of the value of η account for those experimental points lying above the line representing equation (2.8). It could not, however, explain experimental points lying below this line as this would involve negative values of η which would be physically meaningless.

Shaw et al. (1953) reviewed the more important solutions for the shear angle ϕ. From examining a wide range of experimental data they appear to have noted that if η in equation (2.9) could have negative as well as positive values then it would be possible to fit all experimental data. As a negative η is meaningless in terms of the Lee and Shaffer slipline field they attempted to find some other physical interpretation for such an equation. They agreed that equation (2.8) gave the direction of maximum shear stress but contended that this is not necessarily the direction of the shear plane across which the jump in tangential velocity occurs. They reasoned that the coefficient of friction at the tool–chip interface would be dependent on the effective hardness of the chip, and this in turn would depend on the degree of restraint offered by the rigid work behind the shear plane. They concluded that for a given coefficient of friction a certain value of effective hardness would require the shear plane to be at such an angle as to give the required restraint. Thus they modified equation (2.8) to the form

$$\phi = \frac{\pi}{4} + \eta + \alpha - \lambda \qquad (2.12)$$

where η is now the angle between the shear plane direction and the direction of maximum shear stress and can take positive or negative values. Other than when $\eta = 0$ this implies that the directions of maximum shear stress and maximum shear strain-rate no longer coincide. In terms of classical plasticity theory this is only acceptable if the work material is anisotropic. No attempt was made by Shaw et al. to explain the value of η in this way.

In another attempt to obtain better agreement between experiment and theory Kobayashi and Thomsen (1962) have introduced the concept of effectiveness which is essentially a measure of the departure from the minimum energy solution of Merchant (1945). An effectiveness of unity corresponds to equation (2.5) and smaller values of effectiveness give a lower value of ϕ for a given value of $\lambda - \alpha$ than does equation (2.5). By choosing suitable values of effectiveness it is possible to satisfy any experimental value of ϕ. Although useful in collating experimental data, for example, it appears that effectiveness is constant for a given material and cutting speed, the value of the analysis is limited by the lack of any obvious fundamental relationship between effectiveness and work material properties or cutting speed.

There have been many other attempts to obtain shear angle equations and in some cases the same equations or very similar equations have been obtained independently by a number of researchers. Shaw (1984) has reviewed the better known shear angle analyses and has concluded that it is unlikely that a relatively simple shear angle relationship is to be found. Hill (1954) observed that shear plane theories in general gave poor agreement with experiment and questioned the idea that there was a unique solution for ϕ based on the parameters λ and α. In fact he suggested that there was an infinite number of possible steady states based solely on these parameters and that a unique solution could only be found by tracing back to the starting conditions. Still using the shear plane model he determined a permissible range of solutions by taking as acceptable only those solutions which did not give overstressing in the rigid regions adjacent to A and B (Fig. 2.1). In this way he showed that equation (2.5) gave shear angle values outside this range and that equation (2.8) formed part of the upper boundary of the range. From Fig. 2.5 it can be seen that many experimental results fall above Hill's permissible range.

2.3 CURLED CHIP SOLUTIONS

In the shear plane model (Fig. 2.1) the chip velocity is constant across the chip thickness and the chip is straight. This conflicts with the experimental observation that the chip is usually curved and only contacts the tool for a short distance before curling away. Kudo (1965) and Dewhurst (1978) have presented slipline fields to account for this. The main features of these are a curved shear plane to give the required velocity gradient across the chip and a slipline field at the tool–chip interface to represent the plastic deformation which must occur to accommodate contact between a curved chip and a flat tool cutting face.

The slipline field and hodograph for machining with a curled chip proposed by Dewhurst (1978) are given in Fig. 2.6. As pointed out by Dewhurst the hodograph demonstrates that all velocity boundary conditions can be satisfied, namely rigid translation across ABC, rigid rotation across ABDE and zero component of normal velocity across CE. In his analysis Dewhurst initially allowed a small triangular plastic region to exist in the chip adjacent to and above A with one boundary of this region a curved free surface joining the work and chip outer surfaces. The hydrostatic stress could therefore be determined throughout the deforming region starting from the free surface. With this field it was in principle possible to obtain unique solutions for given values of tool rake angle and tool–chip interface shear stress which Dewhurst used as his friction parameter. He investigated this type of solution over a very wide range of realistic geometries using the matrix technique developed by Collins (1968) and Dewhurst and Collins (1973) but could find no solutions in which the force and moment equilibrium conditions imposed by a free chip could be satisfied. He speculated that possibly in cases where the chip is not free as when machining with a chip breaker then such solutions might be found. He later refuted this idea (Dewhurst 1979). In the absence of a free surface and with A now representing the sharp corner where the work and chip surfaces meet as shown in Fig. 2.6, which Dewhurst (1978) concluded was the only type of solution possible, the hydrostatic stress is no longer fixed and the process is not uniquely determined. Noting that neither the uniqueness theorem nor the limit theorems (on which the upper bound method is based) given by Hill (1951) can apply to a process such as machining with its undefined boundaries, Dewhurst, following Hill (1954) determined a permissible range of solutions by selecting as valid only those solutions which did not give overstressing in the rigid regions adjacent to A.

Dewhurst made a number of comparisons between his predicted results and the experimental results of others. From the only experimental observations on chip

Sec. 2.3]   **Curled chip solutions**   33

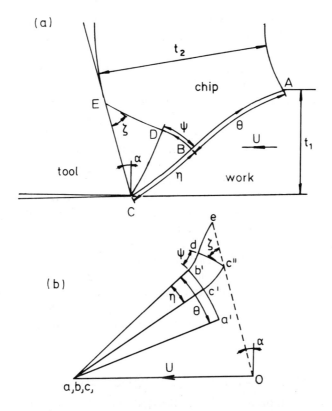

Fig. 2.6 — Curled chip model (Dewhurst 1978): (a) slipline field; (b) hodograph. Method of labelling slipline field and hodograph same as in Fig. 2.4.

curvature which he could find he noted that the measured values fell more or less in the range predicted. This evidence was, however, far from conclusive as only one set of machining conditions was considered. The most widely measured parameter of chip geometry is the shear angle ϕ of the shear plane model (Fig. 2.1). To facilitate a comparison between his predicted results and experimental shear angle results Dewhurst approximated the slipline ABC (Fig. 2.6) by an equivalent shear plane which allowed him to calculate an equivalent shear angle. In the same way as Hill (1954) he found that equation (2.8) formed part of the upper boundary of his permissible range when this was plotted on the basis of ϕ against $\lambda - \alpha$. Consistent with this he showed that along this boundary his field degenerated into the Lee and Shaffer field (Fig. 2.3) with a straight shear plane and no chip curl. Although many experimental shear angle results fall in Dewhurst's permissible range many more fall above it as can be seen, for example, from the results in Fig. 2.5. Dewhurst has contended that in these cases the presence of a built-up edge in the experiments could account for ϕ having been greater than predicted by equation (2.8). However, even when experiments clearly show that there is no built-up edge it is well established that ϕ can be significantly greater than given by this equation. The reason for the discrepancy must therefore apparently be looked for elsewhere.

If the machining process is not uniquely defined then it would be expected as suggested by Hill (1954) that the final steady state for particular cutting conditions will depend to some extent upon the initial conditions. Dewhurst in considering this point could only find two experimental investigations of the influence of starting conditions on the steady state, namely those by Ota et al. (1958) and Low (1962). Unfortunately the results of these investigations are inconclusive with certainly no clear indication that the initial conditions affect the steady state. In particular Low found that the variations in chip thickness ratio (t_1/t_2) resulting from altering the initial conditions were no greater than the variations observed in a parallel set of standard tests. Dewhurst did, however, claim some support from Low's results for the non-uniqueness argument by noting that the variation in t_1/t_2 was 24% for $\alpha = 0°$ while this dropped to 9% for $\alpha = 20°$. This trend agreed with Dewhurst's predicted results which showed a marked decrease in the permissible solution range with increase in rake angle. Some results obtained by Eggleston et al. (1959) for a much wider range of rake angles (5° to 40°) showed this effect even more clearly. In these tests, results were obtained for each rake angle using five different values of undeformed chip thickness. It cannot be ruled out, therefore, that the change in undeformed chip thickness and not the lack of uniqueness resulted in the observed variations, for example, by the change in undeformed chip thickness varying the cutting temperatures which as will be seen later can have a profound effect on chip geometry.

2.4 RESTRICTED TOOL–CHIP CONTACT SOLUTIONS

The benefits of using tools in which the cutting face is cut away so that the tool–chip contact length is less than it would be with a normal tool have long been recognised. One of the earliest and best-known examples of such tools is the Klopstock tool (Klopstock 1925) which by restricting the tool–chip contact length gives reduced cutting forces and improves the wear characteristics of the cutting edge. In modern times the wide use of 'throwaway' tips with groove chip breakers means that many machining operations are carried out under restricted tool–chip contact conditions (Jawahir 1986).

The slipline field and hodograph suggested by Johnson (1962) and Usui and Hoshi (1963) for machining with a restricted contact tool are given in Fig. 2.7. This field is uniquely determined for given values of tool rake angle, restricted contact length, undeformed chip thickness and frictional shear stress along the tool–chip interface. As with the solutions of Lee and Shaffer (1951) AE is a stress-free surface from which the hydrostatic stress can be determined throughout the field. A most interesting feature of the field is that the chip is predicted to flow back into the space created by cutting away part of the tool cutting face. This effect has been widely confirmed experimentally. The equations obtained from the slipline field analysis predict that for given values of tool rake angle α, undeformed chip thickness t_1 and frictional shear stress a reduction in the contact length h will reduce the chip thickness t_2, and in most cases the cutting forces per unit width, and will increase the angle of chip flow η (Fig. 2.7). In an experimental investigation of machining with restricted contact tools, which covered a very wide range of cutting conditions, Jawahir (1986) confirmed all of these predicted trends. However, as with the shear

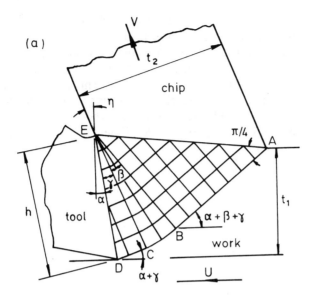

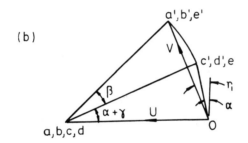

Fig. 2.7 — Restricted tool–chip contact model (Johnson 1962, Usui and Hoshi 1963): (a) slipline field; (b) hodograph. Method of labelling slipline field and hodograph same as in Fig. 2.4.

plane solutions the quantitative agreement between predicted and experimental results was only good over a limited range of conditions. In an attempt to obtain better agreement Jawahir, following Johnson (1962, 1967) and others, proposed a number of alternative slipline fields but only met with limited success.

2.5 LIMITATIONS OF PERFECTLY PLASTIC SOLUTIONS

It is implicit in using the shear plane model and all the other slipline field models described above that the work material is assumed to deform at constant flow stress. Only superficial consideration can therefore be given in analyses based on these models to the actual flow stress characteristics of real work materials which possibly accounts for the generally poor agreement between predicted and experimental results. In general the flow stress of metals varies with strain, strain-rate and temperature and in order to develop an effective machining theory it would seem essential to take account of this. Only in this way might it be expected, for example, that speed and size effects, which experiments show are so important in machining, will be explained. In the following chapters the development of a chip formation model for a material of variable flow stress is described. This is then used as the basis of an approximate machining theory which can account for the influence of work material properties, cutting speed, etc., on the process. The starting point is to consider some slipline field analyses of the experimentally observed flow in chip formation.

3

Slipline field analyses of experimental flow fields

3.1 INTRODUCTION

Once the perfectly plastic assumption is relinquished, and a rigid–plastic hardening material is assumed, the stress equilibrium equations referred to sliplines become (see Appendix A1)

$$\frac{\partial p}{\partial s_1} + 2k \frac{\partial \psi}{\partial s_1} - \frac{\partial k}{\partial s_2} = 0 \quad \text{along a I line}$$
$$\frac{\partial p}{\partial s_2} - 2k \frac{\partial \psi}{\partial s_2} - \frac{\partial k}{\partial s_1} = 0 \quad \text{along a II line} \tag{3.1}$$

where s_1 and s_2 are distances measured along the sliplines. The Geiringer equations (equations (A1.16)) it should be noted remain unchanged because the flow is still assumed to be rigid–plastic. If k varies equations (3.1) show that the variation of p along a slipline depends not only on the angle turned through by the slipline as is the case with equations (2.1) but also on the rate of change of k with distance along the sliplines. Palmer and Oxley (1959), Collins (1979) and Conning et al. (1984) have discussed the nature of rigid–plastic hardening solutions. They have noted that from equations (3.1) the normal derivative of k across a slipline must be continuous. Hence in hardening flows sliplines cannot propagate jumps in velocity (and consequently strain) as are frequently used in rigid–perfectly plastic solutions. In hardening flows velocity discontinuities must, therefore, be replaced by deformation zones in which gradual changes in velocity occur. It follows, as pointed out in Chapter 2, that the shear plane model (Fig. 2.1) and similar models are only valid for a non-hardening material.

So far no analytical solutions for metal working processes have been found once k is allowed to vary even for processes such as extrusion where the rigid–perfectly plastic solutions have long been known. In hardening problems it is generally necessary to determine the distribution of k as part of the solution and as a consequence it is no longer possible to determine simple geometric properties of hardening slipline fields such as those of Hencky for the perfectly plastic case (see Appendix A1). Hardening problems are therefore intrinsically more complicated than non-hardening problems. A notable attempt to obtain solutions to strain-hardening problems has been made by Collins (1979). He has discussed the relation of rigid-hardening solutions to rigid–perfectly plastic solutions and a possible method of deriving the former from the latter using a singular-perturbation technique. Only limited success has so far been achieved with this approach.

Although no analytical solutions to hardening problems have been obtained, slipline fields have been constructed for hardening flows using visioplasticity techniques. The basis of this approach is to use an experimental flow field to determine the velocities of flow which in turn are used to determine the strain-rates and hence the directions of maximum shear strain-rate. The latter are then taken as slipline directions and used in conjunction with known stress boundary conditions to construct a slipline field. The distribution of strain and hence k is then determined by integrating the strain-rate with respect to time along streamlines and a stress analysis carried out using equations (3.1). Slipline fields obtained in this way clearly show the extent to which the velocity discontinuities in rigid–perfectly plastic solutions open up in hardening flows. They also show that the distributions of p calculated from equations (3.1) can be dramatically different from those calculated from equations (2.1). Conning and Oxley (1988) have recently given a comprehensive review of visioplasticity techniques. Their application to machining is now considered.

3.2 EXPERIMENTAL FLOW FIELDS

The objective of the experimental work in the investigations to be described was to obtain a flow field from which the velocities of flow could be determined for a typical plane of flow in chip formation for conditions as near as possible to plane strain, steady-state conditions. There have been two main approaches to obtaining such fields.

In the first, ciné films are taken through a microscope, while machining, of the polished and etched side of a workpiece which, neglecting edge effects, will be a plane of flow and on projection the paths of well-identified grains can be traced to give streamlines of flow in the work, plastic zone and chip. In an early application of this method Palmer and Oxley (1959) checked the effect of side spread on the observed flow by using the velocities in the chip and work measured from the experimental tracings to estimate by continuity the true chip thickness which was then compared with the apparent chip thickness as measured from the projected ciné films. In this way it was found that for rake angles less than 30° the apparent thickness could be as much as 20% too small but for larger rake angles the effect was very much less. It was concluded that the observed fields for larger rake angles could be regarded as representative of the process but that caution must be exercised in using them. If side flow is restricted by a toughened glass plate which still allows pictures of the flow to be taken then these limitations can be largely overcome. With this method care must be taken to ensure that the friction between the glass and deforming metal is low in order to maintain plane strain conditions as closely as possible. Childs (1971) has described such an approach. In the early applications of the ciné filming technique the limitations of the available photographic equipment imposed severe restrictions on the cutting speed. For example, in the experiments of Palmer and Oxley the speed was limited to $U \approx 12.5$ mm/min. Modern equipment removes this limitation and makes possible the application of the technique to practical cutting speeds. The main advantage of this method is that the flow is observed as it actually occurs. This is particularly useful under non-steady-state conditions. For the steady-state conditions of interest in the visioplasticity analyses to be described this advantage is relatively unimportant. Typical examples of flow fields obtained for approximately steady-state conditions using the ciné filming method are given in Figs 3.1 and 3.2. In all cases the work material was low carbon steel and machining was carried out using a planing process. The field in Fig. 3.2 was obtained by Enahoro and Oxley (1966) in an investigation of the flow around the cutting edge which in practice can of course never be perfectly sharp.

In the second method a specimen, split on what will be a plane of flow with a grid scribed or printed on one of the inner surfaces, is used in conjunction with a 'quick-stop' device to freeze the deformed grid so that it corresponds to actual and not stopping conditions. The size of the grid used depends on such factors as the extent of the plastic region, the grain size of the metal and the amount of detail required in the analysis to be carried out on the experimental flow field obtained. Scribing or engraving of the grid has frequently been used. However, when the grid is on an inner surface of a split specimen, which is subsequently clamped in order to give approximately plane strain conditions, then filling in of grooves or modification of the flow field obtained may cause problems. In this connection printed grids using the photoresist method have proved far more effective. Detailed descriptions of this method have been given by Hastings (1967) in relation to machining and by Farmer and Fowle (1979) in relation to extrusion. A vital factor with this method is that once steady-state conditions have been achieved the tool and work should be disengaged as rapidly as possible. Hastings (1967) considered the various mechanical 'quick-stop' devices used in investigating the chip formation process and concluded that they were all deficient in this sense (see section 5.2). In view of this he designed and built a device actuated by a small explosive charge so that the tool was given a very high acceleration and hence a very short retraction time. This explosive 'quick-stop' device which has proved highly effective was later made even more efficient by Bao *et al.* (1976). A typical chequer board grid flow field for orthogonal machining obtained using an explosive 'quick-stop' device is given in Fig. 3.3. (More details of the techniques used in obtaining flow fields of this kind are given in section 5.2.) The work material in this case was a resulphurised low carbon steel. It should be noted that for steady flow the initially horizontal grid lines which are set parallel to the cutting velocity in Fig. 3.3 represent streamlines of flow in the experimental flow field with the distance between the intersecting grid lines giving the speed.

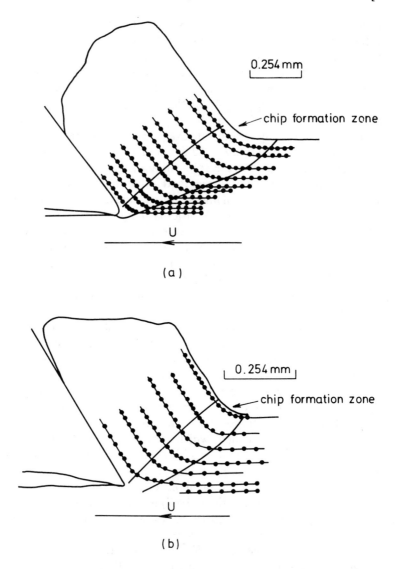

Fig. 3.1 — Experimental flow fields obtained using ciné filming method: (a) $U = 0.0125$ m/min; (b) $U = 15$ m/min.

3.3 GENERAL OBSERVATIONS FROM EXPERIMENTAL FLOW FIELDS

From Figs 3.1 and 3.3 it can be seen that the plastic zone in which the chip is formed is of substantial thickness with the streamlines following smooth curves from the work into the chip and that this is so not only at slow cutting speeds (Fig. 3.1(a)) but also at relatively high cutting speeds (Figs. 3.1(b) and 3.3). Fig. 3.2 shows that a stagnation point of the flow exists on the non-perfectly sharp tool cutting edge with material above the stagnation point flowing into the chip and material below the stagnation point flowing back into the work. From the distance between the experimental points in Fig. 3.2, which are for equal time intervals and hence represent the speed of flow, it

Sec. 3.3] General observations from experimental flow fields

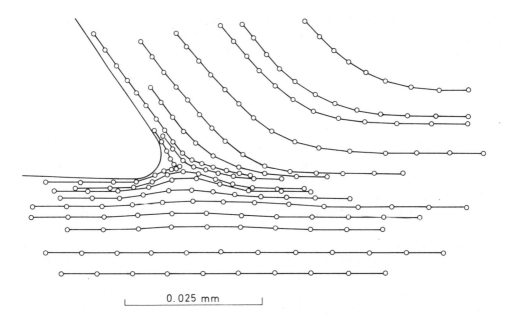

Fig. 3.2 — Experimental flow field obtained using ciné filming method for flow around cutting edge: $U = 0.06$ m/min.

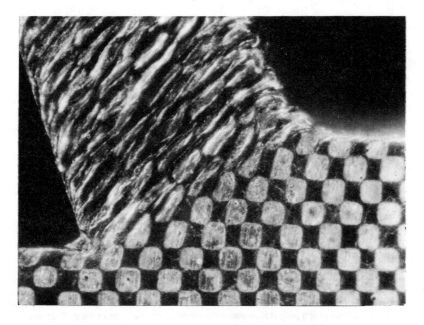

Fig. 3.3 — Experimental flow field (deformed grid) obtained using 'quick-stop' method: $U = 105$ m/min; grid size, ≈ 0.05 mm sided squares.

can be seen that material nearest the tool is retarded relative to the rest of the flow. This means that the chip undersurface and the machined surface will have surface layers which are swept back relative to adjacent material. This can also be seen from the deformed grid in Fig. 3.3.

3.4 SLIPLINE FIELD ANALYSES

The flow fields used in the analyses to be considered were all obtained at very slow cutting speeds ($U \approx 12.5$ mm/min). Variations in flow stress could therefore be assumed to result from strain-hardening alone with strain-rate and temperature effects negligible. In no case in the reported work was it found possible to determine directions of maximum shear strain-rate from the measured velocities with sufficient accuracy to construct slipline fields directly and approximate step-by step methods had to be applied.

In the first investigation of this kind Palmer and Oxley (1959) used an experimental flow field similar to that in Fig. 3.1(a) to determine the boundaries of the plastic zone in which the chip was formed. The positions of these boundaries, represented by a1,a11 and e5,e11 in Fig. 3.4, were estimated by noting where the flow first

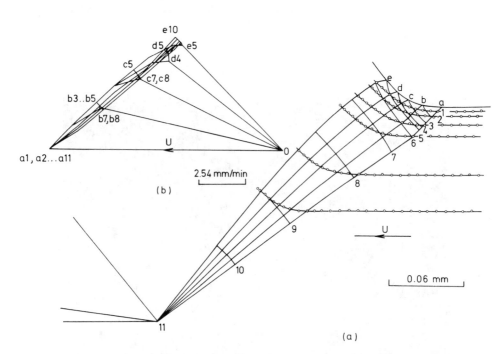

Fig. 3.4 — Palmer and Oxley's (1959) slipline field: (a) slipline field; (b) hodograph. The sliplines and their orthogonal images in the hodograph are labelled a, b, c, etc. (I sliplines) and 1, 2, 3, etc. (II sliplines). The velocity at intersections of sliplines in the slipline field is measured from the hodograph pole (O) to the corresponding intersection in the hodograph.

departed from that of the rigid work and where the flow became that of the rigid chip. No consideration was given to the further plastic deformation which is known to occur at the tool–chip interface (Figs. 3.2 and 3.3) and the plastic zone was terminated in a stress singularity at the tool cutting edge. This was done purely for convenience and it was recognised that in a hardening flow such a singularity was inadmissible. Assuming that a1,a11 and e5,e11 as boundaries between plastically deforming and rigid material would be sliplines (see Appendix A1), two sets of orthogonal curves representing a slipline field were fitted within these boundaries with the sliplines at the free surface making the required angle $\pi/4$ with this surface. The velocity of the work entering the plastic zone was then used as a boundary condition to determine by continuity (by constructing the hodograph — see Appendix A1) the velocities throughout the plastic zone. Adjustments were then made to the assumed field until the corresponding velocities agreed closely with the values measured from the experimental flow field. Fig. 3.4 represents the final slipline field and hodograph found in this way. It is of interest to note that in constructing the hodograph it was found possible to obtain a velocity gradient across e5,e11 consistent with the observed chip curl.

In order to check the slipline field for stress by applying equations (3.1) to determine the distribution of p throughout the field it was first necessary to determine the distribution of k. The method used by Palmer and Oxley was as follows. The components of strain-rate at points along streamlines were calculated from velocities measured from the hodograph and these were then used to determine the corresponding values of effective strain-rate. This was then integrated with respect to time along streamlines to give the distribution of effective natural strain. The corresponding values of effective stress and hence k were then found from a plastic stress strain curve for the work material obtained from a compression test. This procedure is described in more detail in Appendix A2. Having found the distribution of k, equations (3.1) in their finite difference form were used to calculate p by working along sliplines. Starting at the point a1 on the free surface where $p = k$ and is compressive it was found that the field was approximately consistent for stress. In particular, in working round slipline loops such as a6,e6,e7,a7,a6 it was found that the change in p was approximately zero as it should be while clearly this would not have been the case had equations (2.1) been used — see Hencky's first theorem in Appendix A1. Also, the stresses along the boundaries of the plastic zone were reasonably consistent with the measured cutting forces. If the stress analysis was started at points on the free surface other than a1 then less satisfactory agreement was obtained. In fact no satisfactory solution which was consistent for both stress and velocity could be found in the region of the free surface near to e5. It had been noted in the experiments that the surface of the chip was always rough and that this irregular behaviour appeared to originate along the free surface of the plastic zone. In view of this Palmer and Oxley concluded that their lack of success in obtaining a satisfactory solution at the free surface might reflect a real condition, namely cracking as opposed to continuous plastic flow along this surface. To eliminate the unacceptable stress singularity at the cutting edge from their field they postulated the existence of a small free surface in this region and showed how their field might be acceptably terminated at this surface.

A striking demonstration of the importance of allowing for strain-hardening and using equations (3.1) to calculate p was given by Oxley et al. (1961). In a similar investigation to that of Palmer and Oxley they calculated the distribution of p along the work–plastic zone boundary using both equations (2.1) and (3.1). A typical result is given in Fig. 3.5. As drawn in Fig. 3.5 the boundary slipline AB is a I line and as it

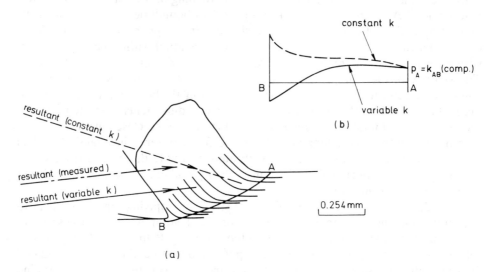

Fig. 3.5 — Hydrostatic stresses along work–plastic zone boundary: (a) flow field and forces; (b) hydrostatic stress distributions.

represents the start of detectable plastic flow it can be assumed that $k = k_{AB}$ is constant along it. Hence, the corresponding equation of equations (3.1) can be integrated to give

$$p + 2k_{AB}\psi - \int \frac{\partial k}{\partial s_2} \, ds_1 = \text{constant along AB} \qquad (3.2)$$

Starting from A where $p = k_{AB}$ (compressive) the distribution of p along AB can be found from equation (3.2). In passing from A to B the curvature term ($2k_{AB}\psi$) gives a compressive increment to p while the strain-hardening term ($\int (\partial k/\partial s_2) ds_1$) gives a tensile increment. In Fig. 3.5 distributions of p along AB are given for k assumed constant and for k increasing into the plastic zone as a result of strain-hardening. It can be assumed that the resultant cutting force is transmitted across AB, and the force directions corresponding to the two stress distributions together with the measured resultant direction are given in Fig. 3.5. It can be seen that excellent agreement exists between the measured direction and that corresponding to the stress distribution calculated from equation (3.2). On the other hand when k is assumed constant the normal force on AB is clearly overestimated. The difference in the hydrostatic stress at B calculated in the two ways can be seen to be of the order of $3k_{AB}$ with equation (3.2) giving a tensile stress in this region. From the position of the resultant force determined from the stresses along the boundary slipline and from the values of shear stress along the tool–chip interface calculated from the experimental forces Oxley *et al.* concluded that tool–chip contact extended down to the cutting edge, thus refuting the idea proposed by Palmer and Oxley of a free surface in this region. The magnitude of the calculated shear stress led them to suggest that the contact was plastic along the interface and that this would be expected if a curved chip was to contact a flat tool cutting face.

A detailed investigation of the flow along the tool–chip interface was made by Enahoro and Oxley (1966). In this they constructed the slipline field and hodograph given in Fig. 3.6 using the given cutting conditions together with experimentally measured values of tool–chip contact length, chip curvature and tool cutting edge radius. The construction was carried out by a trial-and-error process with adjustments made to the slipline field until the hodograph showed that the following velocity boundary conditions were approximately satisfied: (1) material entering the plastic zone from the work and re-entering the work from the plastic zone had to have the velocity of the rigid work (i.e. cutting velocity); (2) material leaving the plastic zone and entering the chip had to have a velocity consistent with the rigid body rotation of the chip; (3) the flow adjacent to the tool face including the radiused cutting edge had to be in a direction parallel to the tool at the point considered. For stress boundary conditions the only constraint placed on the construction was that the shear stress at the tool face should oppose the flow of material. It can be seen from Fig. 3.6 that a stagnation point of the flow occurs at the tool cutting edge. Enahoro and Oxley assumed that the material divides with a very small free surface at this point. For this reason the sliplines were drawn at an angle $\pi/4$ to the surface at the stagnation point. Palmer and Yeo (1963) have considered this part of the flow somewhat differently and have suggested that a small triangular cap of dead (rigid) metal exists in the region adjacent to the stagnation point. Enahoro and Oxley concluded that the assumption of such a cap of metal in their field would only have modified the flow significantly in the immediate region of the stagnation point. A comparison of the streamlines in Fig. 3.6(a), which were constructed from the hodograph, with the experimental streamlines obtained by Enahoro and Oxley for the flow around the cutting edge (Fig. 3.2) shows that the slipline field models the flow from the velocity viewpoint extremely well. By calculating the deformation of an initially square grid as it passed through their slipline field Enahoro and Oxley showed that material near the cutting face was swept back in agreement with experimental observations — see, for example, Fig. 3.3. They pointed out that Zorev (1963) and Wallace and Boothroyd (1964) had concluded that this resulted from the so-called sticking friction effect with the layer of chip material in contact with the tool stationary. They considered that this view was inconsistent with steady-state flow and more akin to built-up edge formation and that as sweeping back of the chip contacting layer was observed under steady-state conditions their own model which accounted for it by a retarded rather than a sticking layer was more acceptable for such conditions. Enahoro and Oxley did not check their field for stress.

In an attempt to construct a slipline field which would take account of both the plastic zone in which the chip is formed and the plastic zone at the tool–chip interface Roth and Oxley (1972) initially proposed a field which in essence consisted of a combination of the fields in Figs. 3.4 and 3.6. They found that although such a field

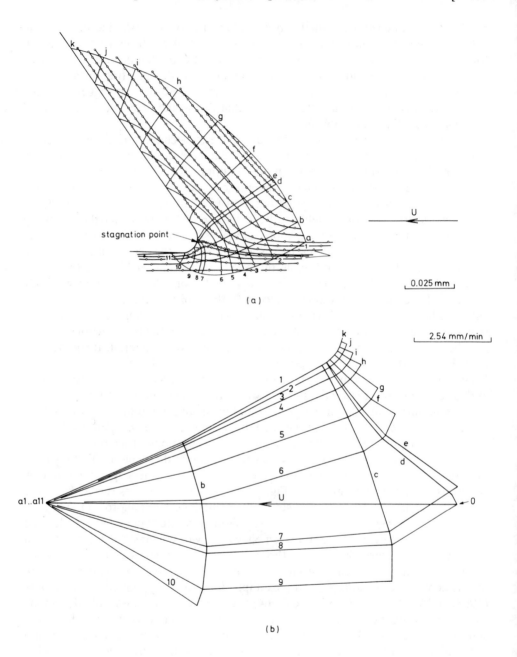

Fig. 3.6 — Enahoro and Oxley's (1966) slipline field for flow along the tool–chip interface: (a) slipline field; (b) hodograph. Method of labelling slipline field and hodograph same as in Fig. 3.4.

could be adjusted to give good agreement for velocities it was in general unsatisfactory for stress. In particular the resultant force calculated from the stress distributions at the tool–chip interface was in very poor agreement with the force calculated

Sec. 3.4] **Slipline field analyses** 47

from the stresses along the work–plastic zone boundary and also with the measured resultant cutting force. They noted that with their field the normal stress at the tool–chip interface increased towards the end of the tool–chip contact and that this was in conflict with distributions of normal stress measured experimentally using photoelastic techniques (see section 7.1) which show that this stress decreases, as would be expected, towards the end of the contact. Usui and Makino (1967) obtained similar results using the same type of field.

Following their lack of success with the combined field, Roth and Oxley proceeded to construct a slipline field using a slow cutting speed chequer board grid flow field as a basis. This, while not enabling the directions of maximum shear strain-rate to be determined with sufficient accuracy to construct a slipline field directly, did give sufficient information for a field to be constructed by trial and error. This information included the approximate position of the plastic zone boundaries and the centre of chip curvature from which the plastic tool–chip contact length was determined by noting where the chip started to curl away from the tool. The final slipline field and hodograph found in this way are given in Fig. 3.7. In constructing

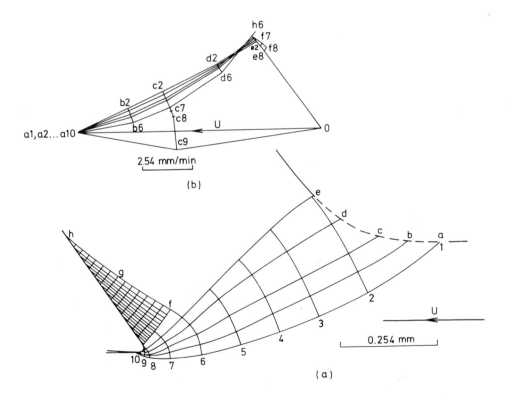

Fig. 3.7 — Roth and Oxley's (1972) slipline field: (a) slipline field; (b) hodograph. Method of labelling slipline field and hodograph same as in Fig. 3.4.

the field it was found that while equations (3.1) had to be used in order to satisfy stresses in the chip formation zone, material at the tool–chip interface had been sufficiently hardened to approximate closely to a perfectly plastic material and in this region a field based on equations (2.1) was applicable. The shear stress and hence the inclination of the sliplines along the tool–chip interface were assumed constant. As in the analysis of Palmer and Oxley (1959) it was found that along the free surface no satisfactory solution could be found and this was again attributed to the observed cracking along this surface. The hodograph in Fig. 3.7 shows that the velocity conditions are all approximately satisfied with the velocities along the boundaries e2,e6 and e6,h6 consistent with the rigid body rotation of the chip and with the flow at the tool–chip interface parallel to the interface. There is a stagnation point of the flow at the tool cutting edge with the flow adjacent to the cutting face retarded relative to the rest of the flow. The field therefore accounts for the deformation of the newly machined surface and the swept-back nature of the chip surface which has been in contact with the tool.

The boundary stresses and forces for the slipline field in Fig. 3.7 are given in Fig. 3.8. It can be seen that the forces transmitted by the work–plastic zone boundary AB

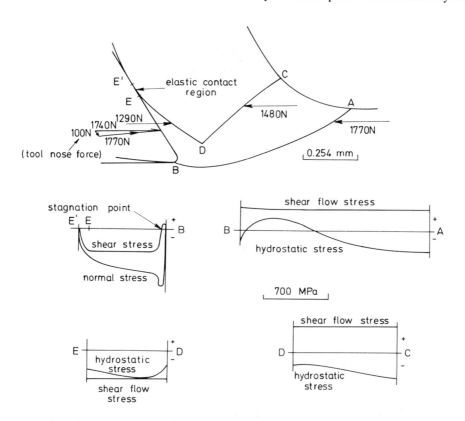

Fig. 3.8 — Boundary stresses and forces for slipline field in Fig. 3.7. Note that shear stresses are shown positive when they exert a clockwise couple on the element on which they act and direct stresses including hydrostatic stresses are shown positive when tensile.

and the tool–chip interface BE′ are in reasonable agreement as are the forces across CD and DE′ as they should be if no forces act on the chip above E′. The hydrostatic stress just in advance of the cutting edge is predicted to be tensile but at the actual edge is compressive. This differs from the result of Palmer and Oxley (1959) who showed this stress to be tensile right at the edge. This is because of the curvature of the boundary slipline near to B which was ignored in the earlier analysis. The normal stress at the tool–chip interface now decreases in moving away from the cutting edge as would be expected. This results from sliplines' such as f6,h6 in Fig. 3.7 curving in the opposite sense to the corresponding sliplines in Fig. 3.6. It is interesting to note the similarity between the tool–chip interface slipline field in Fig. 3.7 and that in Dewhurst's rigid–perfectly plastic solution for a curved chip (Fig. 2.6). Over the part of the tool–chip contact length EE′ (Fig. 3.8) Roth and Oxley assumed that the contact was elastic and showed that a Hertzian stress distribution for this region was consistent with the normal stresses at the interface in the plastic region.

3.5 COMPUTER-AIDED METHODS

The laborious nature of manually constructing and analysing slipline fields in the way described above is obvious. It is therefore of some comfort to report that in more recent work Conning *et al.* (1984) have developed computer-aided methods for constructing slipline fields from experimental flow fields and then carrying out the stress analysis. With these methods smoothing techniques are employed to reduce irregularities in the flow resulting from the granular structure of the metal used and to compensate for errors introduced in tracing and measuring the flow. In this way the directions of maximum shear strain-rate can be determined with far more accuracy and form a good basis for constructing the slipline field. The method has so far been applied mainly to extrusion but should be applicable to machining even allowing for the greater irregularities observed in experimental flow fields for machining compared with those for extrusion.

4
Parallel-sided shear zone theory

4.1 CHIP FORMATION MODEL

In the previous chapter it was shown that the strain-hardening properties of the work material had a profound effect on the hydrostatic stress distribution in the chip formation zone. All of the investigations described were made at very slow cutting speeds and it would be expected that at higher, practical speeds account would also have to be taken in determining hydrostatic stresses of variations in flow stress with strain-rate and temperature. In the first attempt to develop a predictive machining theory for a material of variable flow stress which could take account of such effects Oxley and Welsh (1963) introduced the parallel-sided shear zone model of chip formation shown in Fig. 4.1. In this the shear plane AB of the shear plane model (Fig.

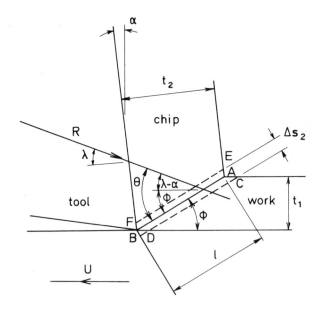

Fig. 4.1 — Parallel-sided shear zone model.

2.1) is assumed to open up so that the boundary CD between the shear zone and work and the boundary EF between the shear zone and chip are parallel to and equidistant from AB and like AB are sliplines. The work velocity is assumed to change to the chip velocity in the shear zone along smooth geometrically identical streamlines with no discontinuities in velocity. The overall geometry of the shear zone model is the same as for the shear plane model with AB and ϕ in Fig. 4.1 geometrically equivalent to the shear plane and shear angle. The force geometry of Fig. 2.2 together with the associated force relations given in equations (2.2) are equally applicable to the shear zone model. However, the resultant force R will not in general pass through the midpoint of AB (Fig. 4.1). Chip curl is again neglected and the overall change in velocity in the shear zone can be represented by the same velocity diagram as in Fig. 2.1 with equations (2.3) giving the corresponding velocities.

4.2 THEORY

For a given rake angle α and undeformed chip thickness t_1 the geometry of the shear zone model (Fig. 4.1) is not completely defined unless the angle ϕ, which will still be referred to as the shear angle, is known. As with the shear plane theories equations are therefore sought for predicting ϕ. The method of analysis is to determine the stresses along AB, in terms of ϕ and work material properties, and from these the magnitude and direction of the resultant force R transmitted by AB; assuming the tool to be perfectly sharp ϕ is then chosen so that the resultant force is consistent with the frictional conditions at the tool–chip interface. From the assumptions made the shear strain will be constant along AB as will the shear strains along CD and EF. The shear strain-rate is assumed to be constant throughout the shear zone and to be given by

$$\dot{\gamma}_{SZ} = \frac{V_S}{\Delta s_2} \qquad (4.1)$$

where V_S is the shear velocity which is equal in magnitude to the velocity discontinuity of the shear plane model (Fig. 2.1) and Δs_2 (Fig. 4.1) is the thickness of the shear zone. If possible temperature variations along AB are now neglected then $k = k_{AB}$ can be assumed constant along AB as can k_{CD} and k_{EF} along CD and EF. AB as drawn in Fig. 4.1 is a I line and noting that it is straight the appropriate equation of equations (3.1), giving the variation in p along AB, reduces to

$$dp = \frac{dk}{ds_2} ds_1 \qquad (4.2)$$

For the assumed model dk/ds_2 can be replaced by the finite difference $\Delta k/\Delta s_2$, where $\Delta k = k_{EF} - k_{CD}$, which will be constant along AB and substituting this and applying equation (4.2) between A and B gives

$$p_A - p_B = \frac{\Delta k}{\Delta s_2} l \qquad (4.3)$$

where p_A and p_B are the hydrostatic (mean compressive) stresses at A and B and $l (l = t_1/\sin\phi)$ is the length of AB. The variation in p along AB is linear with $p_B < p_A$ when, as is usual, $\Delta k/\Delta s_2 > 0$. The normal force acting on AB is therefore given by

$$F_N = \frac{p_A + p_B}{2} lw \qquad (4.4)$$

where w is the width of cut. The shear force on AB is given by

$$F_S = k_{AB} \, l \, w \qquad (4.5)$$

If the angle made by the resultant force R with AB is θ (Fig. 4.1) then

$$\tan\theta = \frac{F_N}{F_S}$$
$$= \frac{p_A + p_B}{2k_{AB}}$$

and substituting for p_B from equation (4.3)

$$\tan\theta = \frac{p_A}{k_{AB}} - \frac{\Delta k}{2k_{AB}} \frac{l}{\Delta s_2}$$

In the previous chapter it was described how in the slipline field analysis of experimental flow fields the hydrostatic stress distribution in the plastic zone could only be determined reliably if the stress analysis was started from a point on the free surface near to where plastic deformation first occurred. In view of this Oxley and Welsh (1963) determined p_A in their analysis by assuming that AB would bend to meet the free surface at the required angle $\pi/4$ in the region where it could still be taken to be parallel to the cutting direction. Assuming that the angle that has to be turned through to meet this condition $(\pi/4 - \phi)$ occurs in negligible distance and noting that at the free surface $p = k_{AB}$, then from the appropriate equation of equations (3.1)

$$\frac{p_A}{k_{AB}} = 1 + 2\left(\frac{\pi}{4} - \phi\right) \qquad (4.6)$$

and substituting for p_A in the previous equation gives

$$\tan\theta = 1 + 2\left(\frac{\pi}{4} - \phi\right) - \frac{\Delta k}{2k_{AB}}\frac{l}{\Delta s_2} \qquad (4.7)$$

θ can also be expressed in terms of the shear angle ϕ, the mean friction angle λ and the rake angle α by the relation

$$\theta = \phi + \lambda - \alpha \qquad (4.8)$$

in the same way as for the shear plane model.

To relate Δk and k_{AB} to the shear flow stress–shear strain curve of the work material Oxley and Welsh assumed a linear stress–strain relation such that

$$\Delta k = m\,\gamma_{EF} \qquad (4.9)$$

where m is the slope of the stress–strain curve and γ_{EF} is the shear strain along EF. γ_{EF} is found by multiplying the strain-rate in the shear zone given by equation (4.1) by the time taken for a particle of material to flow through the shear zone, that is

$$\gamma_{EF} = \frac{V_S}{\Delta s_2}\frac{\Delta s_2}{V_N}$$

where V_N is the velocity normal to AB as given by equations (2.3). Substituting for V_S and V_N from equations (2.3) gives

$$\gamma_{EF} = \frac{\cos\alpha}{\sin\phi\,\cos(\phi - \alpha)} \qquad (4.10)$$

which is the same expression as for the shear strain occurring as material crosses the shear plane. If now it is noted that for the assumptions made half of this strain will have occurred at AB then

$$k_{AB} = k_0 + \tfrac{1}{2}m\,\gamma_{EF} \qquad (4.11)$$

where k_0 is the shear flow stress at zero plastic strain with, for the machining model, $k_0 = k_{CD}$.

For given values of α, λ and t_1 it is now possible to determine ϕ from equations (4.7) to (4.11) if the appropriate values of k_0 and m and the thickness of the shear zone Δs_2 are known. ϕ can then be used together with the corresponding values of k_{AB}, etc., to calculate the cutting forces from equations (2.2).

4.3 EFFECT OF STRAIN-HARDENING ON SHEAR ANGLE

In an early application of the parallel-sided shear zone theory, Oxley and Welsh (1963) attempted to explain the wide variation in experimental shear angle values in Fig. 2.5 in terms of the strain-hardening characteristics of the various work materials used in obtaining these results. Their calculations showed that for the case $m = 0$ (non-hardening material) the relation between ϕ and $\lambda - \alpha$, which can be obtained directly from equations (4.7) and (4.8), closely followed the Ernst and Merchant relation given in equation (2.5) and represented by a straight line in Fig. 2.5, while for hardening materials the predicted results fell below this line. When $m \neq 0$ then it is necessary to know the value of Δs_2 in order to determine ϕ. To find Δs_2 Oxley and Welsh assumed on the basis of the experimental evidence available at the time that the ratio of the length to thickness of the shear zone, $l/\Delta s_2$, would be constant and equal to 10 for the range of cutting conditions and work materials they were considering. With $l/\Delta s_2$ assumed constant, equations (4.7) to (4.11) show that for a given $\lambda - \alpha$ an increase in m/k_{AB} is predicted to decrease ϕ. Oxley and Welsh therefore estimated m/k_{AB} for the materials represented in Fig. 2.5 to see whether this factor could be correlated with the observed variations in ϕ. Fortunately Kobayashi and Thomsen (1959) had provided effective stress–effective natural strain curves for these materials, which are reproduced in Fig. 4.2, but not surprisingly

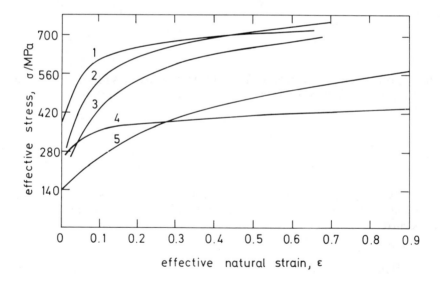

Fig. 4.2 — Effective stress-strain curves: 1, SAE 1112 steel (as received); 2, 2024-T4 aluminium alloy; 3, SAE 1112 steel (annealed); 4, 6061-T6 aluminium alloy; 5, alpha brass.

these were obtained from conventional slow speed (compression) tests. Lacking similar data for the high strain-rates of machining (see next section) Oxley and Welsh used the stress–strain curves in Fig. 4.2 to measure m/k_{AB} for their comparison with the machining results. In effect they assumed that if materials sorted into a certain order on the basis of m/k_{AB} at low strain-rates then they would follow the same order at the high strain-rates applicable to machining. In calculating m/k_{AB} from the curves in Fig. 4.2 they took m as the average slope above an effective strain of 0.2 reasoning from earlier work (Oxley et al. 1961) that this would be more relevant than the initial slope. k_{AB} was taken as the value of k at an effective strain of 0.5, this being assumed to be reasonable on the basis of strain values calculated from equation (4.10). Plane strain and uniaxial conditions were related as described in Appendix A2 with $k = \sigma/\sqrt{3}$ and $\gamma = \sqrt{3}\varepsilon$. The values of m/k_{AB} found in this way are given in Table 4.1.

Table 4.1

Material	m/k
1. SAE 1112 steel (as received)	0.14
2. 2024-T4 aluminium alloy	0.10
3. SAE 1112 steel (annealed)	0.25
4. 6061-T6 aluminium alloy	0.11
5. Alpha brass	0.42

From these results it would be expected that the range of values of ϕ would be lowest for alpha brass which has the highest m/k_{AB} value and then in the order SAE 1112 steel (annealed), SAE 1112 steel (as received), 6061-T6 aluminium alloy and 2024-T4 aluminium alloy. The experimental results for ϕ in Fig. 2.5 can be seen to confirm this trend. Therefore in spite of the somewhat speculative nature of the analysis it does appear that once hardening effects are allowed for it is possible to explain at least in principle why some materials machine with much smaller shear angles than others.

4.4 STRAIN-RATE EFFECTS

From the early experimental work of Kececioglu (1958) and Nakayama (1959) at relatively high cutting speeds and of Palmer and Oxley (1959) at low cutting speeds it appeared that the ratio of the length to the mean thickness of the shear zone ($l/\Delta s_2$) had a value between 6 and 15 for a range of work materials and cutting conditions. Using estimated values of Δs_2 found in this way together with typical values of V_S in equation (4.1) shows that for practical cutting speeds, which are typical say of turning, the average shear strain-rate in the shear zone lies in the range from 10^3 to 10^5/s or even higher. These values are much higher than the strain-rates of 10^{-3} to 10^{-1}/s normally encountered in conventional tension and compression tests. As will be seen, strain-rate can have a marked effect on a material's stress–strain properties with the flow stress of many metals increasing rapidly with increase in strain-rate. Therefore to be realistic any machining theory should take account of strain-rate effects. The temperatures generated in machining and their effect on flow stress would be expected to be equally important but this was not fully recognised in the work now to be considered. Later chapters describe how due account was eventually taken of temperature.

Because of the difficulty of making speeded-up conventional tests to measure the flow stress properties of materials at the high strain-rates appropriate to machining a number of workers have suggested using a machining test itself for this purpose. One of the first investigations of this kind was carried out by Kececioglu (1958). He made turning tests on SAE 1015 steel under approximately plane strain, steady-state conditions in which he measured the thickness of the shear zone (which he approximated to a parallel-sided shear zone), the shear angle and the cutting forces for various values of rake angle, undeformed chip thickness and cutting speed. The shear angle ϕ was calculated from equation (2.11) using measured values of chip thickness t_2. Referring to Fig. 2.2 the forces measured were F_C and F_T and these were used to calculate θ and then F_S from the equations

$$\theta = \phi + \tan^{-1}\left(\frac{F_T}{F_C}\right) \tag{4.12}$$

and

$$F_S = (F_C^2 + F_T^2)^{1/2} \cos\theta \tag{4.13}$$

The thickness of the shear zone Δs_2 (Fig. 4.1) was measured from photomicrographs of 'quick-stop' chip sections. From these it was relatively easy to pick out the boundary between the undeformed work and the shear zone but less easy to pick out the boundary between the deformed chip and shear zone. These measurements had therefore to be treated with some caution. Once ϕ, F_S and Δs_2 had been determined then k_{AB} and $\dot{\gamma}_{SZ}$ were calculated from equations (4.5) and (4.1) using the given values of α, U, t_1 and w. The results obtained showed in agreement with previously reported work at lower strain-rates that flow stress increased with increase in strain-rate.

Oxley (1963b) pointed out that Kececioglu (1958) had taken no account of possible differences in shear strain on AB for the different test conditions used and that this could have masked the effect he was looking for. Oxley overcame this difficulty by using the equations obtained from the parallel-sided shear zone analysis to determine additional information from Kececioglu's test results. From the experimental values of ϕ, θ, Δs_2 and k_{AB}, together with the given cutting conditions, he calculated Δk from equation (4.7) and then m from equations (4.9) and (4.10). He then found k_0 from equation (4.11). Values of k_0 and m, which define the stress–strain curve for a particular strain-rate, derived in this way are shown plotted against shear strain-rate values calculated from equation (4.1) in Fig. 4.3. It can be seen that k_0 increases with increase in strain-rate which is the expected trend. The values of m rather surprisingly fall on two curves, both curves showing a decrease in m with increase in strain-rate. The reason for the points' falling on two curves is not clear. Oxley and Stevenson (1967) pointed out that the top curve represented points for which the shear strain was approximately 2 and the mean temperature in the shear zone was approximately 250°C while the corresponding values for the lower curve were 4 and 350°C. In the light of this they considered possible explanations. They

Sec. 4.4] **Strain-rate effects** 57

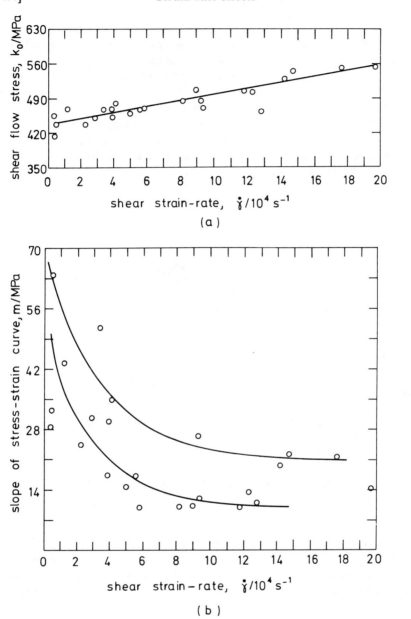

Fig. 4.3 — Values of k_0 and m calculated from machining test results: (a) k_0 results; (b) m results.

quickly ruled out temperature effects as the cause and concluded that the most likely explanation was to do with the assumption of a linear stress–strain relation when in reality compression tests on similar steels to that used in the machining tests showed that experimental points gave a good fit with a relation of the form

$$\sigma = \sigma_1 \varepsilon^n \tag{4.14}$$

where σ and ε are the uniaxial stress and natural strain and σ_1 and n (strain-hardening index) are constants. Equation (4.14) shows that the slope of the stress–strain curve decreases as strain increases and if it were applicable to machining conditions this would explain why for larger strains the use of a linear approximation gives smaller values of m as noted. In view of this Oxley and Stevenson decided to use the machining results to calculate the constants σ_1 and n in equation (4.14) and in this way to determine whether the variation of n with strain-rate was more consistent with expectations than that for m. The method they used is now described.

Equation (4.14) can be represented as shown in Fig. 4.4 with the points CD, AB

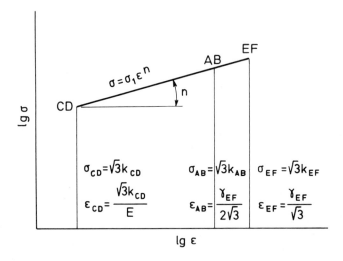

Fig. 4.4 — Construction for obtaining equations (4.15) and (4.16).

and EF indicating the stress and strain on the corresponding sliplines in Fig. 4.1. Plane strain and uniaxial conditions have again been related in the usual way as described in Appendix A2. The strain on CD (zero plastic strain) is taken as the elastic strain just at the point of yielding (note that E is the elastic modulus). From Fig. 4.4

$$n = \frac{\lg(\sqrt{3}k_{EF}) - \lg(\sqrt{3}k_{CD})}{\lg(\gamma_{EF}/\sqrt{3}) - \lg(\sqrt{3}k_{CD}/E)}$$

$$= \lg\left(\frac{k_{EF}}{k_{CD}}\right) \bigg/ \lg\left(\frac{E\gamma_{EF}}{3k_{CD}}\right) \qquad (4.15)$$

and

$$lg(\sqrt{3}k_{EF}) = lg(\sqrt{3}k_{AB}) + n\left[lg\left(\frac{\gamma_{EF}}{\sqrt{3}}\right) - lg\left(\frac{\gamma_{EF}}{2\sqrt{3}}\right)\right]$$

which on simplifying gives

$$n = 3.32 \; lg\left(\frac{k_{EF}}{k_{AB}}\right) \qquad (4.16)$$

For given values of Δk ($\Delta k = k_{EF} - k_{CD}$), k_{AB} and γ_{EF} equations (4.15) and (4.16) can be used to find n and the initial flow stress σ_0 ($\sigma_0 = \sqrt{3}k_{CD}$). σ_1 is then found from equation (4.14). Values of n, σ_1 and σ_0 calculated in this way are shown plotted against $\dot{\varepsilon}$ ($\dot{\varepsilon} = \dot{\gamma}_{SZ}/\sqrt{3}$) in Fig. 4.5. Although there is considerable scatter in the

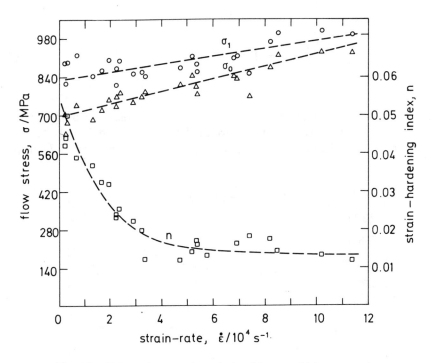

Fig. 4.5 — Values of n, σ_1 and σ_0 calculated from machining test results.

results certain trends can be clearly seen. σ_0 and σ_1 both increase with increase in strain-rate and their difference ($\sigma_1 - \sigma_0$) decreases at higher strain-rates. n decreases with increase in strain-rate in a similar way to m (Fig. 4.3) but the results, unlike those for m, appear to follow one line.

Oxley and Stevenson compared their results with results obtained by Manjoine

(1944) from tension tests for strain-rates in the range from 10^{-6} to 10^3/s. (Manjoine's strain-rates are based on engineering strain (change in length divided by original length) but for the small strains he considered, the difference from natural strain-rate would be negligible.) Manjoine described his test material as a commercial low carbon, open-hearth steel. Although this material was not the same as that used in the machining tests it was similar and Oxley and Stevenson considered it worthwhile to compare the two sets of results. The tension tests were made for various initial temperatures of the test specimens (room temperature, 200, 400 and 600°C) and it was found that the best agreement between machining and tension test results was obtained when the room temperature results were used in making the comparison. In this connection it is interesting to note that the initial work temperature in the machining tests was also room temperature and that in both sets of tests plastic deformation would have led to a temperature rise with this more pronounced at higher strain-rates. Room temperature values of initial flow stress σ_0 and flow stress at a direct strain of approximately 0.19 obtained by Manjoine are given in Fig. 4.6. From these n can be found directly from the relation

$$n = lg\left(\frac{\sigma_{0.19}}{\sigma_0}\right) \bigg/ lg\left(\frac{0.19E}{\sigma_0}\right) \tag{4.17}$$

where the strain at yield is again taken as the elastic strain just at the point of yielding. Once n is known then σ_1 can be determined from equation (4.14). The results obtained from both the machining and tension tests are given in Fig. 4.7. In spite of

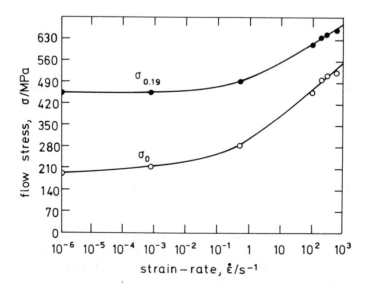

Fig. 4.6 — Manjoine's (1944) tension test results.

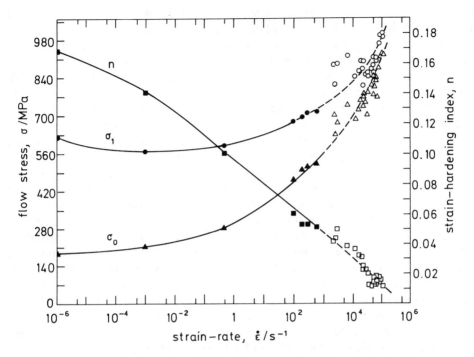

Fig. 4.7 — Values of n, σ_1 and σ_0 calculated from machining and tension test results: open symbols represent machining values and filled-in symbols represent tension test values.

the scatter in the machining results it can be seen that the fit between the two sets of results is remarkably good, thus giving some support to the idea of using machining as a high strain-rate property test.

4.5 PREDICTING THE INFLUENCE OF CUTTING SPEED ON SHEAR ANGLE AND CUTTING FORCES

Following the introduction of strain-rate into the machining analysis a number of investigations were made of the influence of cutting speed on shear angle and cutting forces and these are now described. In all of these the ratio $l/\Delta s_2$ was assumed to be equal to 10. In the investigations considered first the work material flow stress characteristics were represented by the curves in Fig. 4.3 or by similar curves.

For given values of α, λ, U and t_1 the method of calculating ϕ is as follows. A reasonable guess is first made of the value of ϕ and this is used to determine Δs_2 ($\Delta s_2 = t_1/10 \sin \phi$). $\dot{\gamma}_{SZ}$ is then calculated from equation (4.1), with V_S given by equations (2.3), and the corresponding values of k_0 and m are found from Fig. 4.3 with a single average curve used to represent the m values. Then, substituting the value of γ_{EF} given by equation (4.10) in equations (4.9) and (4.11), Δk and k_{AB} are determined and used with equation (4.7) to find θ which in turn is used with equation (4.8) to find λ. In general this value of λ will not agree with the given value and ϕ is suitably adjusted and the process repeated until the difference between the two values of λ is acceptably small. Oxley and Welsh (1963) using this approach showed that their predicted results when plotted on the basis of ϕ against $\lambda - \alpha$ as in Fig. 2.5 indicated that for a given $\lambda - \alpha$ a decrease in cutting speed from 300 to 0.01 m/min gave a decrease in ϕ of approximately 10°. Their results also showed a size effect because of the dependence of $\dot{\gamma}_{SZ}$ on t_1, a decrease in t_1 increasing $\dot{\gamma}_{SZ}$. In this connection for a given $\lambda - \alpha$ a four-fold increase in t_1 was predicted to decrease ϕ by about 5°. Experimental results were presented showing good agreement with these predictions. Oxley and Welsh (1967) applied the theory to a re-analysis of the machining results of Merchant (1945) which he had used to determine the slope s (equation (2.6)) of his proposed linear shear flow stress against normal stress relation. Plotting shear and normal stress values on the shear plane calculated from his machining results Merchant showed that the relation was approximately linear with $s \approx 0.23$ and that if this value was substituted in equation (2.6) then this gave a good fit with his experimental shear angle values. In their analysis Oxley and Welsh started by using Merchant's machining results to calculate m and k_0 as functions of strain-rate in the way described in the last section. These showed reasonably close agreement with the results in Fig. 4.3 with the m results very close to the average m curve and with the k_0 results lying on a line parallel to but about 25% higher than the k_0 line in Fig. 4.3. They then used these values to predict ϕ from the parallel-sided shear zone theory for Merchant's cutting conditions including his experimental values of λ. The line obtained when the predicted results were plotted on a ϕ against $\lambda - \alpha$ graph showed excellent agreement with Merchant's experimental shear angle results. Oxley and Welsh also plotted k_{AB} against $(p_A + p_B)/2$, the average normal stress on AB (Fig. 4.1), using results obtained from their analysis and like Merchant obtained an approximately linear relation of slope ≈ 0.23. They pointed out that their results were a consequence of the hardening properties of the work material, which led to p varying along AB, and that they were in no way indicative of a Bridgman-type relationship in which the shear stress is directly affected by the normal stress as assumed by Merchant and for which a slope of 0.23 would be unacceptably large.

In the applications of the parallel-sided shear zone theory so far considered λ has been used as the tool–chip interface friction parameter in the same way as for the shear plane solutions. If λ could be measured by some independent friction test or if its value did not vary greatly with cutting conditions this would not be a great disadvantage. However, it has only been found possible to measure λ by machining tests and the variations in λ with α, U and t_1 even for a given work material-tool material combination are too great to allow the use of an average value. Thus λ must be measured from machining tests before 'predictions' can be made, which is clearly unsatisfactory. This has led Oxley (1966) and others including Rowe and Spick (1967) to suggest that the shear strength (shear flow stress) in the chip material adjacent to the tool–chip interface might be a more effective friction parameter. This idea is based on experimental observations, such as those described in the previous chapter, which show that plastic deformation occurs in a layer of the chip adjacent to the surface which slides over the tool. This can be contrasted with normal metallic sliding friction conditions where plastic deformation is limited to the tips of contacting asperities and it can usually be assumed that for a given sliding pair and lubrication condition the coefficient of friction μ, and hence λ, will be more or less constant. The nature of tool–chip interface friction is considered in detail in section 7.3. For the present, attention will be limited to the replacement of λ by the average shear stress at the interface as the friction parameter.

The average shear stress at the tool–chip interface is given by

$$\tau = \frac{F}{hw} \tag{4.18}$$

where F (Fig. 2.2) is the frictional force at the interface, h is the tool–chip contact length and w is the width of cut. Oxley (1966) has obtained an expression for h in the following way. Taking moments of the normal stress on AB (Fig. 4.1) about B it can be shown that the resultant force R (Fig. 2.2) cuts the tool–chip interface at a distance

$$x = \frac{2t_1 \sin \theta}{\cos \lambda \sin \phi} \frac{\tfrac{1}{3}p_A + \tfrac{1}{6}p_B}{p_A + p_B}$$

above the cutting edge, where p_A and p_B are the hydrostatic stresses at A and B. If now the distribution of normal stress along the tool–chip interface is known then h can be found. For example if, as assumed by Oxley, the distribution is triangular with the maximum value at B then $x = h/3$ and therefore

$$h = \frac{t_1 \sin \theta}{\cos \lambda \sin \phi} \frac{2p_A + p_B}{p_A + p_B} \tag{4.19}$$

If λ is replaced by τ as the friction parameter the method of calculation is the same as before up to the point where λ is found from equation (4.8). Then, F is found from equations (2.2) and h from equation (4.19), with p_A and p_B given by equations (4.6) and (4.3), and these are substituted in equation (4.18) to give τ. The process is repeated with ϕ suitably adjusted until the calculated and given values of τ are in reasonable agreement. Working backwards from machining results Oxley (1966) used the above equations to determine τ and showed that this was reasonably constant and equal to about half the value of k_{AB} for the conditions he considered.

Fenton and Oxley (1967) using τ as the friction parameter investigated the influence of cutting speed on the cutting forces F_C and F_T (Fig. 2.2) over a very large range of speeds (30 to 3000 m/min).† They represented the work material flow stress characteristics using equation (4.14) with the required values of σ_1 and n found from Fig. 4.5. While recognising that τ would most likely vary with cutting speed, etc., they assumed for the purpose of their analysis that it could be taken to be constant and equal to approximately half the value of k_{AB}. Their results showed that over the entire speed range considered ϕ was predicted to increase with increase in speed while the tool–chip contact length h and force normal to the cutting velocity F_T were

---

† It should be noted that, even at a cutting speed of 6000 m/min, F_C is increased by less than 3% by the force resulting from the rate of change of momentum in chip formation. However, as speed is increased further this force increases rapidly.

predicted to decrease. The force in the cutting velocity direction F_C was predicted to decrease up to a speed of approximately 300 m/min and then to increase rapidly with further increase in speed. They explained this result by noting that up to 300 m/min the reduction in F_C was caused by the increase in ϕ which reduced the area of the plane AB and hence the force F_S which is given by the product of k_{AB} and the area of AB (see equation 4.5). However, at higher cutting speeds although ϕ still increased with increase in speed the reduction in area of AB was more than compensated by the increase in k_{AB} resulting from the increase in strain-rate, and F_S, and as a consequence F_C, no longer decreased but actually increased. Fenton and Oxley considered that their results cast great doubt on the suggestion made by a number of workers that superhigh cutting speeds could lead to a dramatic fall in F_C. They recognised, however, that their analysis had limitations particularly those resulting from the lack of consideration given to temperature.

Fenton and Oxley (1968–1969) in a later investigation used the parallel-sided shear zone theory to calculate τ from the machining results of Merchant (1945). They found that τ varied markedly (a variation of approximately 60% from the lowest to the highest value) with α, U and t_1 and that this variation could only be explained if account was taken of the strain-rates and temperatures in the plastic zone at the tool–chip interface. They therefore suggested that τ should be expressed in terms of strain-rate and temperature and showed how this might be done using the velocity-modified temperature concept of MacGregor and Fisher (1946) which combines strain-rate and temperature in a single parameter. This approach is considered in detail in later chapters. Assuming a linear relation between τ and velocity-modified temperature to represent the tool–chip interface frictional conditions, Fenton and Oxley (1969–1970) made predictions of ϕ, F_C, F_T and λ (which it will be noted is now a derived result) for a very wide range of cutting conditions. Their analysis was based on the experimental machining results for SAE 1112 steel obtained by Eggleston *et al.* (1959) and Kobayashi and Thomsen (1960). The approach used was to determine the work material flow stress parameters σ_1, n and τ from a small number of machining test results using the machining theory in reverse and then to apply these in making predictions for a much wider range of cutting conditions. Working in this way they predicted results which were in good agreement with experimental results over very wide ranges of α (5° to 35°), U (0.03 to 300 m/min) and t_1 (0.05 to 0.254 mm). In particular, their predicted results showed in agreement with experimental results that ϕ increased with increase in α, t_1 and U and that F_C and F_T decreased with increase in α and with increase in U. λ was predicted to decrease with increase in t_1 and with decrease in α and the experimental results largely confirmed this.

5
Experimental investigation of the influence of speed and scale on the strain-rate in the chip formation zone

5.1 INTRODUCTION

In applying the parallel-sided shear zone model (Fig. 4.1) described in the previous chapter it was assumed that the thickness of the shear zone Δs_2 would be proportional to the length l of this zone. In other words that Δs_2 would depend on scale only and not as might be expected on cutting speed. The main investigations on which this assumption is based (Kececioglu 1958, Nakayama 1959) are open to criticism because of the methods used in measuring the size of the shear zone. For example, in the most comprehensive experiments, which are those made by Kececioglu, this zone was measured by using a mechanical, spring-activated, 'quick-stop' device to freeze the chip sections which were then mounted, polished and etched. In this way the boundary between the shear zone (deformed material) and work (undeformed material) could be picked out reasonably well but the boundary between the shear zone and chip (deformed material) was far more difficult to identify. Also, the 'quick-stop' device used allowed cutting to continue for a distance of the same order as the undeformed chip thickness during the retraction of the tool which could have allowed changes to occur during stopping. Kececioglu's results show some decrease in Δs_2 with increase in cutting speed but in view of the method of measuring Δs_2 this result must be treated with caution.

Stevenson and Oxley (1969–1970) largely overcame the above difficulties by making orthogonal machining experiments using an explosive 'quick-stop' device together with a printed grid to measure the deformation in the chip formation zone. They used the results obtained to determine the influence of cutting speed and undeformed chip thickness on the size of the chip formation zone and the strain-rates within this zone. A detailed description of this investigation is now given.

5.2 EXPERIMENT

The explosive 'quick-stop' device used in the experiments was basically the same as that developed by Hastings (1967) and referred to in Chapter 3. In this a block which holds the tool is located in vertical guides and rests on a shear pin. On detonation the explosion exerts pressure through a piston on the top of the tool-block. The stress on the shear pin therefore builds up and at a certain stress the shear pin fails and the tool is accelerated away from the work. Hastings had originally used hardened steel shear pins but Stevenson and Oxley found that cast iron shear pins resulted in a more consistent type of failure in the shear pin and more consistency in the time between firing the explosive charge and failure of the pin. Hastings measured the average acceleration of the tool block in the initial 13 mm of its movement using a high speed ciné camera, obtaining a value of 15×10^5 m/s^2. If this acceleration is uniform throughout the movement then at a cutting speed of 240 m/min the cutting action would only continue for a distance of about 0.005 mm during the relative deceleration of the tool and work. However, measurement of the tool-block acceleration over the first 0.05 mm by Stevenson and Oxley using a wire strain gauge glued between the body of the 'quick-stop' device and the tool-block gave an acceleration of 2.5×10^5 m/s^2 and hence a deceleration distance of 0.03 mm at 240 m/min. That this was approximately correct was shown by slight changes in the undeformed chip thickness over this distance in specimens obtained at this speed. No figure for acceleration over the initial travel distance is available for Kececioglu's device. Ciné picture measurements, however, show the acceleration of the explosive device to be about ten times greater. Also, as Hastings has shown, the explosive device causes far less distortion of the chip. A flow field which is typical of those obtained by Stevenson and Oxley is given in Fig. 3.3. It can be seen that the chip thickness remains uniform, the rake angle is maintained and the undeformed chip thickness is uniform up to approximately 0.025 mm before the end of the cut.

The work material used in the experiments was a resulphurized low carbon steel, CS 1114 (Australian standard) of chemical composition 0.13% C, 1.4% Mn, 0.25% S, 0.019% P. This was chosen because it gave a continuous chip, with very little or no built-up edge over a wide range of cutting conditions. Discs of this material were parted off from bright bar of 0.11 m diameter and turned down to 0.10 m diameter and fine ground to a width of approximately 3.2 mm. In a test three such discs were clamped tightly together on a mandrel, in order to obtain approximately plane strain conditions on the centre disc. The experimental machining arrangement is given in Fig. 5.1. With this the diameter of the discs was reduced by a straight-edged cutting tool which was fed radially inwards. The cutting edge was set normal to the cutting and feed directions. The cutting tools used were brazed carbide recessing tools with rake angles of 10°, 20° and 30° and a clearance angle of 6°. The cutting face was lapped on a diamond lapping machine before each test. The experiments covered a wide range of cutting speeds (5 to 250 m/min) and undeformed chip thicknesses (0.127 to 0.274 mm) with the latter equal to the radial feed measured in mm per revolution of the workpiece. The lathe used in the tests was a Churchill–Denhams SR-10V with a 35 h.p. variable speed motor and spindle speeds infinitely variable from 15 to 1500 rev/min.

The centre disc was polished on one side and thinly plated with copper. The copper was then etched to produce a grid using the photographic process described

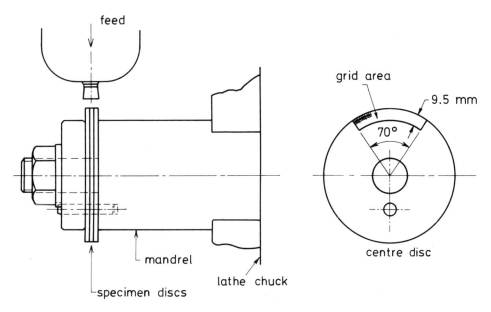

Fig. 5.1 — Experimental machining arrangement.

by Hastings (1967). The radial grid master from which the grids were printed on the specimens was prepared in the following way. A segment of the grid was drawn up in ink 50 times larger than the required size. This segment was then reduced to one-fifth size on a lithographic camera, five identical segments being made. These were then butted together to form about 70° of arc. This composite was then reduced to one-tenth size on a microfilm camera, making a total reduction of 50. The depth of grid segment obtained in this way was approximately 9.5 mm. The resulting image, on 35 mm film, was printed on a glass plate. To this plate was glued a locating spigot, concentric with the grid to allow location on the specimen for contact printing. It was considered most convenient not to make a complete grid circle on the specimen but to concentrate attention on the 70° segment to get good quality. This required that the 'quick-stop' device would have to be 'fired' at a rotational position which, after the short delay involved, would cause the cutting action to be stopped about half-way through the grid. A photo-cell was used to sense the rotational position of the specimen for this purpose. To control the total reduction in disc diameter before 'quick-stopping' a microswitch actuated by the in-feed was provided in series with the photo-cell. The 9.5 mm depth of grid segment allowed about four tests to be performed on each specimen, with about 10 revolutions of cutting for each test.

5.3 ANALYSIS OF EXPERIMENTAL FLOW FIELDS

Although the grid boundaries of the experimental flow fields obtained, such as that given in Fig. 3.3, are actually arcs and radial lines it can be seen that when analysing the small area of the deformation zone it is reasonable to assume the arcs to be straight lines and the sides of the "squares" to be equal. Likewise it is reasonable to treat the process as if it were straight line cutting with the cutting velocity constant across the undeformed chip thickness and equal to the disc circumference multiplied by the number of revolutions per minute of the workpiece. Fig. 5.2 gives a tracing of

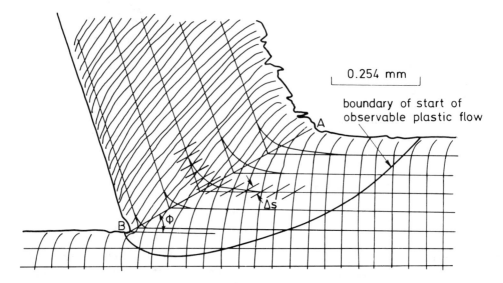

Fig. 5.2 — Tracing of deformed grid given in Fig. 3.3: $\alpha = 20°$; $U = 105$ m/min; $t_1 = 0.264$ mm.

the deformed grid in Fig. 3.3 with the grid lines which were originally parallel to the work velocity represented by smooth curves which average out the irregularities actually occurring in these lines in the chip formation zone and chip. Fig. 5.2 is typical of the traced flow fields obtained from the experiments and the analysis of the deformation will be described by reference to it.

In principle the strain-rates in the chip formation zone can be determined by measuring the velocities of flow and hence the velocity gradients from experimental flow fields such as that in Fig. 5.2. However, as with the slipline field analyses described in Chapter 3 it was found that the velocities could not be measured with sufficient accuracy for this purpose. In view of this an approximate method of calculating strain-rates based on the parallel-sided shear zone model of chip formation (Fig. 4.1) was developed. With this method the direction of maximum shear strain-rate in the chip formation zone is assumed to be constant as with the parallel-sided shear zone model and is determined in the same way as with this model. Referring to Fig. 5.2, a straight line AB has been drawn which passes through (approximately) the intersections of the streamlines in the work and chip (assumed parallel to the tool cutting face) and this is taken as the direction of maximum shear strain-rate within the chip formation zone. The velocity diagram in Fig. 2.1 can now be used to represent the velocity changes in the chip formation zone (Fig. 5.2) with

V_S the total change in shear velocity, i.e. in the direction AB, in the zone. With the shear plane (Fig. 2.1) and parallel-sided shear zone (Fig. 4.1) models the velocity diagram (Fig. 2.1) is automatically satisfied but this is not generally true for the measured chip and work velocities and angle ϕ (i.e. the angle between AB and the work velocity) in Fig. 5.2. Because of errors in construction and measurements and deviations from the idealised model used, e.g. some chip curl, the velocity diagram cannot satisfy both measured velocities and ϕ simultaneously but this can easily be overcome by changing any one of these parameters by a small amount (less than 10%). For most of the cutting conditions considered the velocity diagram giving the best fit with the velocities in the chip formation zone measured from the deformed grid was found to be that defined by ϕ and the chip velocity V together with the given rake angle α. The velocity diagram defined in this way (Fig. 5.3) was therefore used in the strain-rate analysis.

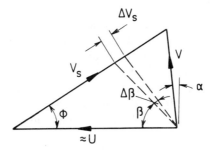

Fig. 5.3 — Velocity diagram used in strain-rate analysis.

Considering the centre streamline in the chip formation zone (Fig. 5.2), lines parallel to AB representing directions of maximum shear strain-rate have been drawn at approximately equal time intervals along the streamline (i.e. through the corners of the grid) and hence approximately equal distances Δs apart, where Δs is measured normal to AB. The change in shear velocity ΔV_S (i.e. in the direction AB) across an element of thickness Δs can be represented as shown by the broken lines in the velocity diagram given in Fig. 5.3 where β is the angle turned through in reaching the element, measured from the work velocity, and $\Delta \beta$ is the further angle turned through in crossing the element. The mean value of the maximum shear strain-rate in the element is given by

$$\dot{\gamma}_{max} = \frac{\Delta V_S}{\Delta s} \tag{5.1}$$

From the geometry of Fig. 5.3, ΔV_S can be expressed in terms of the chip velocity V, the angle ϕ and the angles β and $\Delta\beta$. Doing this and substituting in equation (5.1) gives

$$\dot{\gamma}_{max} = \frac{V\cos(\phi - \alpha)\sin(\Delta\beta)}{\Delta s \sin(\phi + \beta)\sin(\phi + \beta + \Delta\beta)} \quad (5.2)$$

The strain occurring across the element $\Delta\gamma$ can be found by multiplying $\dot{\gamma}_{max}$ by the time a particle takes to flow across the element, which gives

$$\Delta\gamma = \frac{\sin(\Delta\beta)}{\sin(\phi + \beta + \Delta\beta)\sin(\phi + \beta)} \quad (5.3)$$

Equations (5.2) and (5.3) are now sufficient to calculate from an experimental flow field the strain-rates and strains in the chip formation zone.

5.4 RESULTS AND DISCUSSION

In making calculations the procedure was as follows. A transparency of an experimental flow field was projected onto a drawing board to give an approximately 300 times magnification. The actual magnification in each case was found by measuring an interval on the undeformed grid with a toolmaker's microscope. The deformed grid boundaries were drawn to give the streamlines and points along them representing approximately equal time intervals. The velocity of the chip V was found by the change in lengths of the grids along the streamlines and the angle ϕ found from the construction in Fig. 5.2. For equal time intervals (i.e. equal distances Δs measured normal to AB) along a streamline the angles β and $\Delta\beta$ were measured. These were then used with the corresponding values of V, ϕ, α and Δs to determine $\dot{\gamma}_{max}$ from equation (5.2) and $\Delta\gamma$ from equation (5.3). With the larger values of undeformed chip thickness, β and $\Delta\beta$ were measured using three streamlines near the centre of the chip formation zone, these all giving approximately the same results, but for smaller values of undeformed chip thickness this had to be reduced to one or two streamlines near the centre of the zone.

The distributions of maximum shear strain-rate in the chip formation zone found in this way for four different cutting speeds are given in Fig. 5.4. These results have been plotted taking AB (Fig. 5.2) as the origin of the horizontal scale with the units of this scale equal to increments of Δs. In the tests ϕ did not vary greatly and Δs was therefore very nearly constant ($\Delta s \approx 0.023$ mm) in all tests. The strain-rate can be seen to increase with increase in cutting speed (Fig. 5.4) as would be expected and to have a maximum value which occurs in the region defined by the plane AB (Fig. 5.2). The maximum values of $\dot{\gamma}_{max}$ are plotted against shear velocity V_S (found from the velocity diagram constructed from the measured values of V and ϕ and the given value of α) in Fig. 5.5. Shear velocity is used as it might be expected that this is the velocity to which $\dot{\gamma}_{max}$ is most likely to be related. The results show (Fig. 5.5) that the maximum value of $\dot{\gamma}_{max}$ increases with increase in V_S but that the undeformed chip thickness also influences strain-rate, a decrease in t_1 increasing the strain-rate. This is also shown in Fig. 5.6 where the experimental distributions of $\dot{\gamma}_{max}$ are given for two values of t_1. In Fig. 5.7 the maximum values of $\dot{\gamma}_{max}$ have been plotted against V_S/t_1 and it can be seen that the results are less scattered than when V_S is used as the

Sec. 5.4] Results and discussion

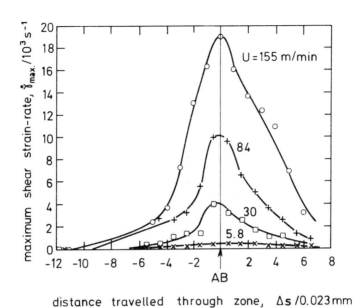

Fig. 5.4 — Distribution of $\dot{\gamma}_{max}$ in chip formation zone showing influence of cutting speed: $\alpha = 20°$; $t_1 = 0.264$ mm.

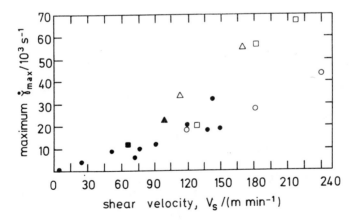

Fig. 5.5 — Variation of maximum $\dot{\gamma}_{max}$ values with shear velocity: $\triangle, \square, \bigcirc$ represent results for $\alpha = 10°$ and $t_1 = 0.127, 0.175, 0.274$ mm respectively; $\blacktriangle, \blacksquare, \bullet$ represent results for $\alpha = 20°$ and the same values of t_1.

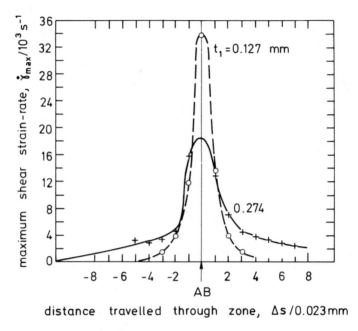

Fig. 5.6 — Distribution of $\dot{\gamma}_{max}$ in chip formation zone showing influence of undeformed chip thickness: $\alpha = 10°$; $U = 120$ m/min.

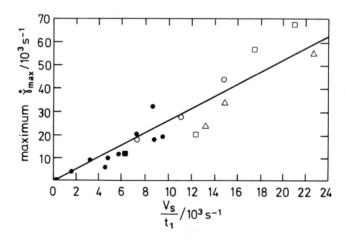

Fig. 5.7 — Variation of maximum $\dot{\gamma}_{max}$ values with shear velocity divided by undeformed chip thickness: symbols same as in Fig. 5.5.

horizontal scale (Fig. 5.5). From the results in Fig. 5.7 the line of best fit passing through the origin is

$$\max(\dot{\gamma}_{\max}) = 2.59 \frac{V_S}{t_1} \tag{5.4}$$

where the units of V_S/t_1 are s^{-1}.

From Figs 5.4 and 5.6 it can be seen that plastic deformation starts well in advance of AB (Fig. 5.2) and continues well beyond it. The chip formation zone is therefore of substantial thickness with the thickness appearing to change only slightly with cutting speed (Fig. 5.4) but significantly with undeformed chip thickness (Fig. 5.6). From a detailed analysis of all of their experimental results Stevenson and Oxley (1969–1970) concluded that the average thickness of the chip formation zone measured over its centre portion changed little with speed but a great deal with undeformed chip thickness, thus supporting the assumption of the parallel-sided shear zone theory that Δs_2 (Fig. 4.1) would depend on scale but not on speed. The results do in fact show that the ratio of the length of the chip formation zone to its thickness is approximately constant for the range of cutting conditions considered with the average value of this ratio equal to 2.6 which can be compared to the values of between 8 and 15 obtained by Kececioglu (1958) and the constant value of 10 assumed in the parallel-sided shear zone analysis. Although supporting the assumption that $l/\Delta s_2$ is constant the results (Figs. 5.4 and 5.6) clearly show that the strain-rate is not constant in the chip formation zone as it is assumed to be in the parallel-sided shear zone theory.

The experimental strain distributions obtained by Stevenson and Oxley confirm the assumption made in the parallel-sided shear zone analysis that half the total strain occurring in the chip formation zone has occurred by the time material reaches AB (Fig. 5.2). The results also show that the values of total strain in the chip formation zone calculated from the experimental flow fields are close to the values determined from equation (4.10) using the corresponding experimental values of ϕ and the given values of α.

Similar results to those in Figs. 5.4 and 5.6 have been obtained by Goriani and Kobayashi (1967) and Crookall and Richardson (1969). These, however, show that the strain-rate increases towards the cutting edge which is consistent with the increase in curvature of streamlines near the cutting edge shown in Fig. 5.2. Stevenson and Oxley (1969–1970) accepted that the strain-rate would be greater near the cutting edge and referred to the work of Enahoro and Oxley (1966) which showed that a stagnation point of the flow existed at the cutting edge (see Chapter 3) which would suggest extremely high strain-rates in this region. In spite of this they contended that their results, such as those in Figs. 5.4 and 5.6, would be representative of the flow over most of the chip formation zone.

6

Work material properties: the influence of strain-rate and temperature

6.1 FLOW STRESS DATA DETERMINED FROM MACHINING TEST RESULTS

The strain-rate distributions in the chip formation zone given in Figs. 5.4 and 5.6 cast doubt on the method of calculating σ_1 and n used by Oxley and Stevenson (1967) which is based on the parallel-sided shear zone model and is described in section 4.4. In this the shear flow stress k_{AB} and the change of shear flow stress across the shear zone Δk are assumed to correspond to the average shear strain-rate in the shear zone given by equation (4.1). This is clearly in conflict with the results in Figs. 5.4 and 5.6 which show that the strain-rate in the region of AB (Fig. 5.2) is much greater than at the entry and exit boundaries of the chip formation zone. To allow for this in calculating σ_1 and n Stevenson and Oxley (1970–1971) made an analysis similar to the parallel sided shear zone analysis but in which the strain-rate on AB (Fig. 5.2) was determined from an equation of the form of equation (5.4) instead of from equation (4.1). This analysis is now described.

AB as established in Fig. 5.2 is assumed to be a direction of maximum shear strain-rate and hence maximum shear stress with the shear strain-rate constant along AB and given by

$$\dot{\gamma}_{AB} = C' \frac{V_S}{t_1} \qquad (6.1)$$

where, for the results in Chapter 5, $C' = 2.59$. It is implicit in using this equation that the maximum values of $\dot{\gamma}_{max}$ in Figs. 5.4 and 5.6 are assumed to occur at AB which for the given results appears reasonable. The shear strain is also assumed to be constant along AB and, from the results described in Chapter 5, to be given by

Sec. 6.1]   Flow stress data determined from machining test results   75

$$\gamma_{AB} = \frac{1}{2} \frac{\cos \alpha}{\sin \phi \cos(\phi - \alpha)} \qquad (6.2)$$

The tool is again assumed to be perfectly sharp with AB transmitting the resultant cutting force. Possible temperature variations along AB are again neglected and $k = k_{AB}$ is assumed constant along AB. Following these assumptions a similar analysis to that carried out in Chapter 4 for the parallel-sided shear zone model, which yielded equation (4.7), can be made. In the new analysis the hydrostatic stress at A (Fig. 5.2) is determined in the same way as for the parallel-sided shear zone model, i.e. from equation (4.6). However, the rate of change of k normal to AB can now be related to the actual strain-rate at AB and it is possible to replace $\Delta k/\Delta s_2$ in equation (4.7) by dk/ds_2 determined in the following way.

In general k will vary with strain, strain-rate and temperature and if history effects are neglected so that $k = f(\gamma, \dot{\gamma}, T)$† where γ, $\dot{\gamma}$ and T are the maximum shear strain, maximum shear strain-rate and temperature then

$$\frac{\partial k}{\partial s_{1,2}} = \frac{\partial k}{\partial \gamma}\frac{\partial \gamma}{\partial s_{1,2}} + \frac{\partial k}{\partial \dot{\gamma}}\frac{\partial \dot{\gamma}}{\partial s_{1,2}} + \frac{\partial k}{\partial T}\frac{\partial T}{\partial s_{1,2}} \qquad (6.3)$$

In determining dk/ds_2 at AB from equation (6.3) for substitution in equation (4.7) the strain-rate term can be neglected as strain-rate can, on the basis of experimental results (Figs. 5.4 and 5.6), be assumed to pass through a maximum at AB. The temperature term is also neglected because of the difficulties involved in calculating the temperature gradient across AB but as will be seen this should not lead to large errors. Therefore

$$\frac{dk}{ds_2} = \frac{dk}{d\gamma}\frac{d\gamma}{ds_2}$$

which can be written

$$\frac{dk}{ds_2} = \frac{dk}{d\gamma}\frac{d\gamma}{dt}\frac{dt}{ds_2} \qquad (6.4)$$

where t is time. If now the flow stress properties of the work material are represented by equation (4.14), that is $\sigma = \sigma_1 \varepsilon^n$ then at AB

---

† An implicit assumption in the treatment of material flow stress properties throughout this book is that an equation of state exists for the materials considered. That is, for a particular material it is assumed that the flow stress will be unique for given values of strain, strain-rate and temperature. Thus possible variations resulting from the paths by which the strain, strain-rate and temperature are reached, which experiments show can be significant, are neglected. It might reasonably be expected that neglect of such history effects, which at this stage would appear prohibitively difficult to allow for, will not introduce serious errors particularly in the case of the steady-state machining process to which attention has mainly been limited.

$$d\sigma/d\varepsilon = n\sigma_{AB}/\varepsilon_{AB} \quad \text{or} \quad dk/d\gamma = nk_{AB}/\gamma_{AB} \tag{6.5}$$

where γ_{AB} is given by equation (6.2). The second term on the right-hand side of equation (6.4) is simply the shear strain-rate given by equation (6.1) and the third term is the reciprocal of the velocity normal to AB which from equations (2.3) can be taken as

$$dt/ds_2 = 1/U\sin\phi \tag{6.6}$$

Substituting from equations (6.1), (6.2), (6.5) and (6.6) in equation (6.4) and noting from equations (2.3) that

$$V_S = U\cos\alpha/\cos(\phi - \alpha)$$

the following relation is obtained:

$$dk/ds_2 = 2C'nk_{AB}/t_1 \tag{6.7}$$

and substituting this for $\Delta k/\Delta s_2$ in equation (4.7) and noting that $l = t/\sin\phi$ (4.7) gives

$$\tan\theta = 1 + 2\left(\frac{\pi}{4} - \phi\right) - \frac{C'n}{\sin\phi} \tag{6.8}$$

Experimentally determined values of θ, ϕ and C' can now be used to calculate n from equation (6.8). The values thus obtained will correspond to the actual strain-rate at AB as will values of k_{AB} determined from the resolved forces along AB. A further advantage of using equation (6.8) is that in calculating σ_1 and n it is no longer necessary to assume a value of strain at the boundary representing the start of plastic flow as it was in the parallel-sided shear zone analysis — see Fig. 4.4. If it is again assumed that for each test condition (each strain-rate) the derived values of n, k_{AB} and γ_{AB} define a single stress–strain curve of the form given in equation (4.14) then σ_1 in this equation is given by

$$\sigma_1 = \frac{\sigma_{AB}}{\varepsilon_{AB}^n} = \frac{\sqrt{3}k_{AB}}{(\gamma_{AB}/\sqrt{3})^n} \tag{6.9}$$

Stevenson and Oxley (1970–1971) have calculated σ_1 and n in the above way using the machining test results considered in the previous chapter which also included cutting forces measured during the tests. They had intended at first to calculate σ_1 and n using values of ϕ and strain-rate found from the 'quick-stop' flow fields together with the forces measured at the instant of 'quick-stop'. However, the variations in these measured cutting forces, strain-rates, etc., caused partly by

Sec. 6.1]   **Flow stress data determined from machining test results**   77

experimental error, but which also reflect real variability of the properties of the work material within the very small test specimen (chip formation zone), tended to obscure the trends. It was therefore decided to use average force and shear angle values (Figs. 6.1 and 6.2) found over relatively long cuts at a number of speeds

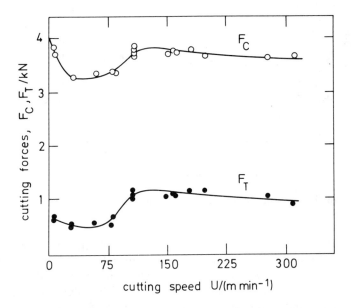

Fig. 6.1 — Experimental cutting forces: $\alpha = 20°$; $t_1 = 0.264$ mm; $w = 9.5$ mm.

keeping α and t_1 constant and to determine the corresponding strain-rates from equation (6.1). The forces were measured using a dynamometer which was a conventional arrangement of semiconductor strain gauges bonded to a tube integral with the explosive 'quick-stop' tool holder. The dynamometer was calibrated while in its operating position on the lathe with dead weights acting through lever arms and push rods which bore directly on the tool edge. The shear angle values were calculated from equation (2.11) using experimentally measured chip thickness values which were determined for each cutting condition from about 20 measurements along a representative length of chip. Further smoothing of the experimental results was achieved by working with values obtained from the curves drawn through the experimental points in Figs. 6.1 and 6.2. In this way Stevenson and Oxley averaged out variations in work material properties and effectively obtained results which were representative of larger specimens.

In their calculations Stevenson and Oxley determined θ, F_S and k_{AB} from equations (4.12), (4.13) and (4.5) using experimental values of ϕ, F_C and F_T. (Note that in determining k_{AB} from equation (4.5) l was taken equal to $t_1/\sin \phi$.) Equations (6.8) and (6.9) were then used to find n and σ_1 with the corresponding strain-rates given by equation (6.1) with C' taken as 2.59. Values of σ_1 and n obtained in this way

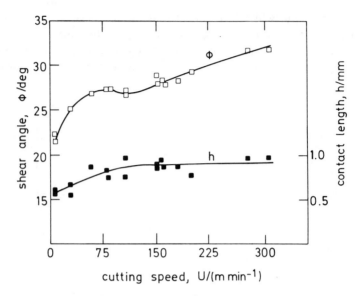

Fig. 6.2 — Experimental shear angles and tool–chip contact lengths: $\alpha = 10°$; $t_1 = 0.264$ mm.

are shown plotted against strain-rate in Fig. 6.3, where $\dot{\varepsilon} = \dot{\varepsilon}_{AB} = \dot{\gamma}_{AB}/\sqrt{3}$. Also given in Fig. 6.3 are values of σ_1 and n, for the same material as used by Stevenson and Oxley, determined from slow speed compression tests and calculated as above from machining results obtained at very low cutting speeds. The values of σ_1 and n found from the machining and compression tests can be seen to fit together very well and show similar trends with strain-rate to the results in Fig. 4.7. The new results should be far more reliable and do, indeed, show far less scatter than the earlier ones (Fig. 4.7) owing to the improvements made in both the experiments and analysis and also because the same material was used for the complete range of tests. As before it was possible to represent σ_1 and n by single curves over the entire range of strain-rate although, as will be seen, the temperature in the tests (along AB in the machining tests) varied from room temperature to approximately 200°C. However, it will be noted that there is a dip and subsequent rise in the σ_1 curve between strain-rates of 10^2/s and 10^4/s and in explaining this it is found that consideration must be given to temperature as well as strain-rate. Methods for calculating temperatures in machining are now considered.

6.2 TEMPERATURE CALCULATION METHODS

Heat is generated in machining by the work done in the plastic zone in which the chip is formed and by friction at the tool–chip interface. In this connection it should be noted that for the shear plane model the total work rate $F_C U$ is equal to $F_S V_S + FV$ and that for the assumptions made this also applies to the parallel-sided shear zone model and to the chip formation model given in Fig. 5.2 with $F_S V_S$ the work rate in the chip formation zone and FV the work rate at the tool–chip interface. If the tool is

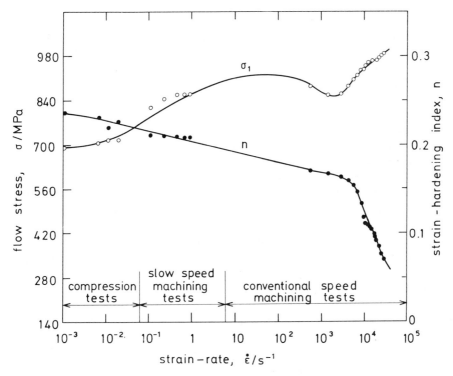

Fig. 6.3 — Values of n and σ_1 calculated from machining and compression test results.

not perfectly sharp, e.g. if wear has occurred on the clearance face, then additional heat will be generated by friction between the wear scar and the newly machined surface.

There have been a number of attempts to calculate temperatures in machining using the shear plane model. In these, and in fact in all of the temperature calculation methods to be considered, it is assumed that all of the work done in chip formation is converted into heat with only a negligible amount of energy being retained within the deformed metal. Experimental work supporting this assumption has been reported by Taylor and Quinney (1934, 1937) and by Bever et al. (1953). Hahn (1951) calculated the shear plane temperature by assuming the shear plane to be a uniform band source moving obliquely through an infinite workpiece. Leone (1954) and Lowen and Shaw (1954) also assumed the shear plane to be a uniform band source but considered that it moved over a semi-infinite workpiece with the proportions of shear plane heat entering the work and chip determined by applying Blok's (1938) partition principle. Trigger and Chao (1951) and Lowen and Shaw (1954) both used Blok's partition principle to calculate the average tool–chip interface temperature assuming that the heat developed at the tool–chip interface was uniformly distributed. Weiner (1955) obtained a solution for the shear plane temperature distribution by assuming that the chip velocity was perpendicular to the shear plane and that the heat conduction in the directions of motion of the work and chip could be neglected. Using relaxation techniques Rapier (1954) calculated temperature distributions in the work, chip and tool which he treated as three separate systems. He also assumed a constant shear plane temperature and a plane uniform heat source at the tool–chip interface with all the interface heat flowing into the chip and none into the tool. Dutt and Brewer (1964) improved the analysis by treating the work, chip and tool as one system, but after making some approximations found that they were able to dispense with the tool region altogether. In this way they were able to determine the proportions of shear plane heat entering the work and the chip, and the proportions of tool-chip interface heat entering the chip and tool. Chao and Trigger (1955) improved their earlier analytical solution for the interface temperature distribution by allowing the fraction of interface heat flowing into the tool to vary along the interface although still assuming a uniform heat source. In their calculations they made use of an iterative analytical procedure involving a grid of real and fictitious point heat sources.

The chief disadvantages of the above methods of temperature calculation result mainly from the simplifications made in the shear plane model. This assumes a velocity discontinuity across the shear plane while in real materials as clearly shown in Chapters 3 and 5 the transition from the work to chip velocity occurs gradually over a finite plastic zone. Also, the velocity of the chip material adjacent to the tool–chip interface is less than the chip velocity, resulting in the characteristic deformation usually observed in this region. It has been assumed in all the cases considered that the generated heat is confined uniformly within the shear plane and at the tool–chip interface rather than being spread over finite plastic zones, and as shown by Boothroyd (1963) this will certainly result in the temperatures being overestimated. Methods of calculating temperatures which are based on Boothroyd's work and which have been widely applied by the author and his co-workers are now described.

The temperature rise in the plastic zone in which the chip is formed is found by considering the plastic work done in this zone and is given by

$$\Delta T_{SZ} = \frac{1-\beta}{\rho S t_1 w} \frac{F_S \cos \alpha}{\cos(\phi - \alpha)} \qquad (6.10)$$

where ρ is the density of the work material, S its specific heat and β is the proportion of heat conducted into the work. There have been a number of attempts to predict β theoretically including that by Weiner (1955) based on the shear plane model of chip formation. Weiner's results are represented by the line in Fig. 6.4 which shows β plotted against $R_T \tan \phi$ where R_T is a non-dimensional thermal number given by

$$R_T = \rho S U t_1 / K \qquad (6.11)$$

with K the thermal conductivity of the work material. Experimental values of β are also given in Fig. 6.4. These were obtained by Nakayama (1956) using a thermocouple technique to measure the heat carried away by the work and by Boothroyd

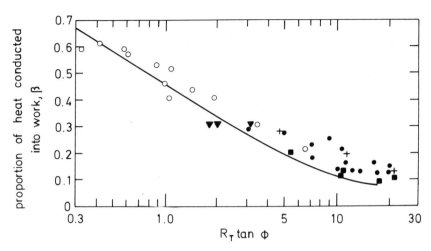

Fig. 6.4 — Theoretical and experimental results for β: line represents Weiner's (1955) theoretical results; ○ represents Nakayma's (1956) experimental results; ▼, ■, ● represent Boothroyd's (1963) experimental results, + represents results calculated by Tay et al. (1974) using finite element method.

(1963) who measured temperature distributions in the work, chip, deformation zones and tool using an infrared radiation method and calculated β from the temperature distribution in the work. (Details of the different methods of measuring temperatures in machining have been given by Boothroyd (1965), Trent (1977) and Shaw (1984).) From the experimental results in Fig. 6.4, it can be seen that Weiner's solution underestimates β. In view of this β has usually been estimated in the work to be described from the following empirical equations which have been derived on the basis of the experimental results in Fig. 6.4:

$$\beta = 0.5 - 0.35 \, lg(R_T \tan \phi) \quad \text{for} \quad 0.04 \leqslant R_T \tan \phi \leqslant 10.0$$
$$\beta = 0.3 - 0.15 \, lg(R_T \tan \phi) \quad \text{for} \quad R_T \tan \phi > 10.0 \tag{6.12}$$

The further limits that β should not exceed 1 or be less than 0 are also imposed. The average temperature along AB (Fig. 5.2) is taken to be given by

$$T_{AB} = T_W + \eta \Delta T_{SZ} \tag{6.13}$$

where T_W is the initial work temperature and η $(0 < \eta \leqslant 1)$ is a factor which allows for the fact that not all of the plastic work of chip formation has occurred at AB. The average temperature at the tool–chip interface which will help determine the average flow stress at the interface is assumed to be given by

$$T_{int} = T_W + \Delta T_{SZ} + \psi \Delta T_M \tag{6.14}$$

where ΔT_M is the maximum temperature rise in the chip which occurs at the interface and ψ ($0 < \psi \leq 1$) is a factor which accounts for the possible variation of temperature along the interface. Boothroyd (1963) has calculated ΔT_M using numerical methods for both triangular (maximum thickness at the cutting edge) and rectangular plastic zones (heat sources) at the interface with no sliding at the interface and hence with all of the frictional work dissipated in these zones. The results of these calculations are represented by the lines in Fig. 6.5. The experimental results given in Fig. 6.5 for the

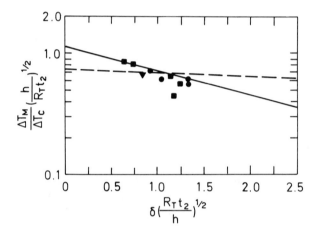

Fig. 6.5 — Calculated and experimental tool–chip interface temperature results: full line represents numerical results assuming rectangular plastic zone and broken line represents numerical results assuming triangular plastic zone; symbols represent experimental results.

ratio of ΔT_M to ΔT_C, the average temperature rise in the chip, were obtained by Boothroyd from his experimentally measured temperature distributions. It can be seen that the calculations based on the rectangular plastic zone give the best fit with experiment. This model has therefore been favoured in the work to be described. If the thickness of the rectangular plastic zone is taken as δt_2, where δ is the ratio of this thickness to the chip thickness, then, as shown by Stevenson (1970), Boothroyd's calculated results for this case (full line in Fig. 6.5) can be represented by the equation

$$lg\left(\frac{\Delta T_M}{\Delta T_C}\right) = 0.06 - 0.195\delta\left(\frac{R_T t_2}{h}\right)^{1/2} + 0.5\,lg\left(\frac{R_T t_2}{h}\right) \qquad (6.15)$$

where ΔT_C is the average temperature rise in the chip which is assumed to be given by

Sec. 6.2] Temperature calculation methods 83

$$\Delta T_C = F\sin\phi/\rho s t_1 w \cos(\phi - \alpha) \quad (6.16)$$

and h is the tool-chip contact length.

Many of the restrictive assumptions associated with the temperature calculation methods reported so far were overcome by Tay *et al.* (1974) who applied the finite element method to determine temperature distributions using the experimental flow fields, from which the distributions of velocity and strain-rate could be found, and associated forces obtained by Stevenson and Oxley (1969–1970) as a basis. In this work, temperature distributions were obtained numerically by solving the steady two-dimensional energy equation using the finite element method with the distributions of heat sources in the chip formation and tool-chip interface plastic zones calculated from the strain-rate and flow stress distributions in these zones. The flow stress data used were those obtained by Stevenson and Oxley (1970–1971, 1973) from machining test results. In considering the interface, account was taken of both the sliding at the interface and the deformation in the plastic zone adjacent to the interface with the associated velocities and strain-rates measured from the experimental flow fields. In the calculations flow stress was taken to be a function of strain, strain-rate and temperature. The chip, work and tool were treated as one system, with account taken of the actual shape and size of the tool, and material properties such as specific heat and thermal conductivity were considered as functions of temperature. Results were obtained for three experimental flow fields and these showed that the temperature along AB, with AB found as in Fig. 5.2, was fairly constant for most of its length but increased rapidly near to the cutting edge. However, as pointed out by Tay *et al.* (1974), the strain-rate also increases in this region and this would to some extent compensate for the temperature rise and help justify the assumption, made in the machining analysis in the previous section, that k_{AB} is constant along AB. The results also allow the temperature gradient normal to AB to be determined and hence an indication to be given of the error introduced by neglecting the temperature term in equation (6.3) when obtaining an expression for dk/ds_2. For the conditions considered the temperature term is about 25% the magnitude of the strain term. Neglect of this term is, however, less important than it might at first appear to be. This follows from the fact that in any test such as a high speed compression test, the results of which might be used to compare machining results with, there will also be a temperature–time gradient similar to that in machining. This would therefore to some extent compensate for neglect of the temperature term in the machining analysis. Tay *et al.* (1974) also used their results to calculate the proportion of heat conducted into the work. These values are indicated by the crosses in Fig. 6.4 and can be seen to fit in well with the experimental results for β.

Tay *et al.* (1976) noted the laborious nature of the analysis carried out by Tay *et al.* (1974) and the substantial computation time involved and concluded that in determining temperatures for a wide range of cutting conditions or in determining temperatures for use with a predictive machining theory a simpler method was needed which did not need an experimental flow field to start with. They achieved this by using a model in which the streamlines in the chip formation zone were assumed to be hyperbolic with the velocities along streamlines given by the diagram in Fig. 5.3. The hyperbola used was of a form which gave strain-rate distributions similar to those found experimentally in Figs. 5.4 and 5.6 with the strain-rate passing through a maximum at AB. The constant in the streamline equation was selected so that the corresponding strain-rates at AB were consistent with those given by equation (6.1). In this way Tay *et al.* (1976) were able to determine the strain-rate distribution in the chip formation zone and neglecting variations in flow stress within the zone assumed that the heat generation at any point would be proportional to the strain-rate at that point. In modelling the tool-chip interface flow account was taken of both sliding and plastic deformation. The sliding velocity was given by an equation based on measurements made from experimental flow fields. The interface plastic zone was taken to extend over the full contact length and to be triangular in shape with maximum thickness at the cutting edge. The shear stress was assumed to be constant along the interface. For the chip formation model developed in this way it was shown that the velocities and strain-rates could be determined for given values of U and α if ϕ (Fig. 5.2) was known. Temperatures could then be calculated given the cutting forces from which the values of average flow stress in the chip formation zone and at the interface, needed in the calculations, could be determined. The method of determining temperature distributions using the finite element method was essentially the same as that used by Tay *et al.* (1974). Account was again taken of the shape and thermal properties of the tool and of the influence of temperature on the work material thermal properties. In the calculations the thickness of the chip formation zone and the maximum thickness of the tool-chip interface plastic zone had to be known and these were taken as $l/2.6$ and $t_2/20$ respectively on the basis of measurements made from experimental flow fields.

Tay *et al.* (1976) used their method to calculate temperature distributions from experimental cutting force and shear angle data obtained when machining S1016 steel (0.19% C, 0.16% Mn, 0.016% P, 0.027% S) using a Sandvik S6 grade carbide tool. In the calculations it was necessary to know the tool-chip contact length. This was found from the equation

$$h = \frac{t_1 \sin\theta}{\sin\phi \cos(\theta + \alpha - \phi)} \qquad (6.17)$$

using experimental values of ϕ and θ. This equation is derived by drawing a line through point A in Fig. 4.1 parallel to R and assuming that no force is transmitted above the point where the line cuts the tool face, i.e. that the distance from the cutting edge to this point is equal to h. Average values of T_{AB} and T_{int} obtained by numerical integration of temperatures along AB and the interface found by Tay *et al.* are given in Fig. 6.6. Also given in Fig. 6.6 are the values of T_{AB} and T_{int} calculated, using the same experimental machining results and work material thermal properties, from equations (6.13) and (6.14) taking the temperature factors η and ψ as unity. In these calculations δ in equation (6.15) was taken as 0.05 which meant that the thickness of the rectangular tool-chip interface plastic zone was assumed equal to the maximum thickness of the triangular plastic zone used by Tay *et al.* The tool-chip contact length was again calculated from equation (6.17). Stevenson *et al.* (1983) using a similar approach to that of Tay *et al.* (1976) calculated temperature

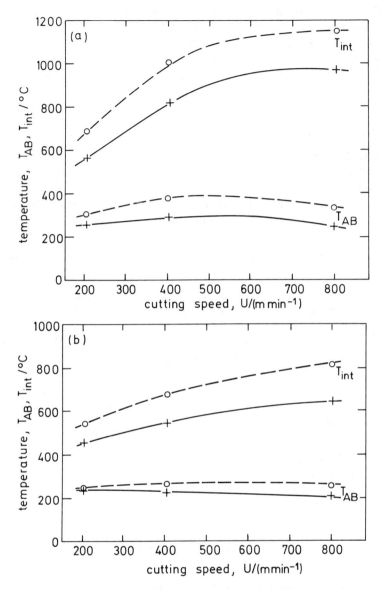

Fig. 6.6 — Comparison of temperatures calculated from equations (6.13) and (6.14) with those calculated by Tay et al. (1976): broken lines represent results given by equations (6.13) and (6.14) with $\eta = \psi = 1$ and full lines represent results obtained by Tay et al.: (a) $\alpha = 5°$, $t_1 = 0.264$ mm; (b) $\alpha = 10°$, $t_1 = 0.127$ mm.

distributions and compared their results with tool temperature distributions measured experimentally using the technique developed by Wright and Trent (1973). In this the change in temper of the high speed steel tool used is measured and correlated with temperature. Good agreement was shown between calculated and experimental results. If on this basis the results of Tay et al. in Fig. 6.6 are accepted as realistic then the use of the factors η and ψ in equations (6.13) and (6.14) to give better estimates of T_{AB} and T_{int} is clearly justified. From the results in Fig. 6.6 it can be seen that η and ψ would need to have values falling in the range 0.7 to 0.95 in order to make the values of T_{AB} and T_{int} found from equations (6.13) and (6.14) equal to the values found by Tay et al.

6.3 THERMAL PROPERTIES

In the applications of the machining theory described in the following chapters attention is mainly limited to plain carbon steel work materials and in calculating temperatures the appropriate temperature-dependent thermal properties have been determined in the following way. The influence of carbon content on specific heat S is found to be small and from data given by Woolman and Mottram (1964) the equation

$$S/(J/(kg\,K)) = 420 + 0.504 T/°C \qquad (6.18)$$

can be used for all of the steels considered. However, there is a marked influence of carbon content on thermal conductivity K and allowance must be made for this and the influence of other elements on K. Hastings (1975) has shown how this can be achieved using experimental results for thermal conductivity given by Woolman and Mottram (1964). In this approach two equations are used which are derived from the experimental results. In the first equation the thermal conductivity at 0°C is expressed in terms of chemical composition. This is then used with the second equation which relates the variation in thermal conductivity to temperature. The relevant equations found from the data of Woolman and Mottram are

$$K_0 = 1/(5.8 + 1.6[C] + 4.1[Si] + 1.4[Mn] + 5[P] + [Ni] + 0.6[Cr] + 0.6[Mo]) \qquad (6.19)$$

and

$$K = 418.68[0.065 + (K_0 - 0.065)(1.0033 - 11.095 \times 10^{-4} T)] \qquad (6.20)$$

where K_0 is the thermal conductivity at 0°C, K is the thermal conductivity at T°C with the units of thermal conductivity, W/(m K), and [C], [Si], etc., are the percentages of the various elements present in the work material considered. Equations (6.19) and (6.20) yield equations of the form

$$K/(W/(m\,K)) = 54.17 - 0.0298 T/°C \qquad (6.21)$$

which is for a steel of chemical composition 0.20% C, 0.15% Si, 0.015% S, 0.72% Mn, 0.015% A*l*, and

$$K/(W/(m\,K)) = 52.61 - 0.0281\,T/°C \qquad (6.22)$$

which is for a steel of chemical composition 0.38% C, 0.1% Si, 0.77% Mn, 0.015% P.

In the temperature calculations the effect of temperature on work material density ρ, which will be negligibly small, has been ignored and ρ has been taken as 7862 kg/m^3 for all of the steels considered.

6.4 INTERPRETATION OF FLOW STRESS DATA IN TERMS OF STRAIN-RATE AND TEMPERATURE

Values of T_{AB} calculated by Stevenson and Oxley (1970–1971) from equations (6.10) to (6.13), using the same experimental machining results (Figs. 6.1 and 6.2) as used in determining σ_1 and n in Fig. 6.3, are given in Fig. 6.7. In these calculations η was

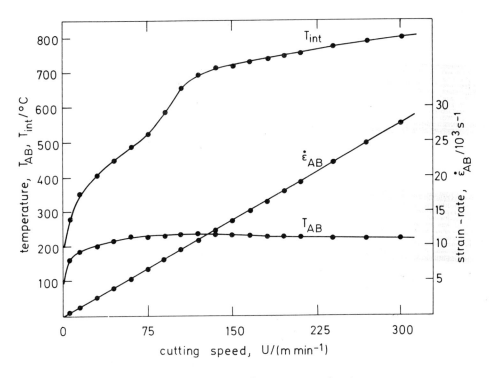

Fig. 6.7 — Calculated values of temperature and strain-rate.

taken as unity and T_W as 22°C. An iterative calculation procedure was necessary as S and K were both assumed to vary with temperature as described in the previous section and to have values corresponding to a temperature $T = T_{AB}$. Also given in Fig. 6.7 are values of $\dot{\varepsilon}_{AB}$ where $\dot{\varepsilon}_{AB} = \dot{\gamma}_{AB}/\sqrt{3}$ with $\dot{\gamma}_{AB}$ found from equation (6.1) taking $C' = 2.59$. From the results in Fig. 6.7 it can be seen that T_{AB} rises rapidly with speed at slow cutting speeds. This is because β, the proportion of heat conducted into the work, is also changing rapidly. At higher speeds, the temperature rise tends to become adiabatic with little heat conducted into the work and T_{AB} becomes approximately constant. The strain-rate $\dot{\varepsilon}_{AB}$ increases approximately linearly with cutting speed over the entire range. Following Stevenson and Oxley it is now possible to explain the dip and subsequent rise in the σ_1 curve in Fig. 6.3 in terms of these temperature and strain-rate results by referring to suitable flow stress data.

Some experimental results of Ohmori and Yoshinaga (1966, 1968) showing the variation of lower yield stress with temperature and strain-rate for a mild steel of similar carbon content (0.14%) and static lower yield stress (182 MPa) to those used in Stevenson and Oxley's tests are given in Fig. 6.8. These results were obtained from

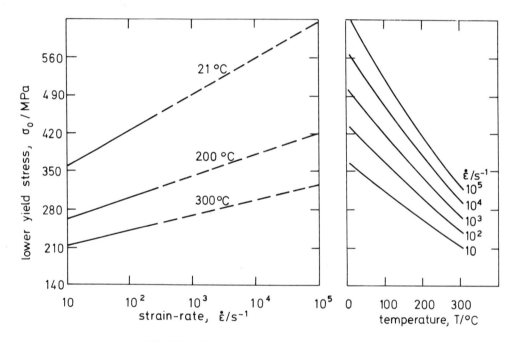

Fig. 6.8 — Experimental lower yield stress results.

high-speed tension tests and cover temperatures up to 300°C (these are starting temperatures and no account has been taken of the small temperature rise in the tests). The tension tests went to strain-rates of up to 3.7×10^2/s, which is somewhat lower than the minimum value in the conventional speed machining tests (Fig. 6.3). From the results in Fig. 6.8 it can be seen that, for the ranges of strain-rate and temperature considered, the yield stress increases linearly with the logarithm of the strain-rate and decreases approximately linearly with temperature. The broken lines in Fig. 6.8 are extrapolations of the tension test results to strain-rates of the order met in the machining tests. The right-hand side of Fig. 6.8 was constructed for the purpose of interpolation to any temperature between 0°C and 300°C.

In Fig. 6.9 a curve of lower yield stress has been constructed from the data in Fig.

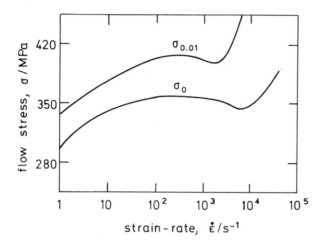

Fig. 6.9 — Comparison of machining and tension test results.

6.8 using the values of T_{AB} and $\dot{\varepsilon}_{AB}$ given in Fig. 6.7. This curve is of the same form as that for σ_1 in Fig. 6.3 and in particular shows the same dip and subsequent rise in stress. However, the σ_1 curve is higher than the lower yield stress curve, which is not surprising as the strain corresponding to σ_1 ($\varepsilon = 1$) is far greater than the strain at initial yield. To make a more direct comparison it was assumed that the values of σ_1 and n (Fig. 6.3) found from the machining test results could be used to represent the stress–strain curve for the corresponding values of temperature and strain-rate over a sufficient range to allow equation (4.14), i.e. $\sigma = \sigma_1 \varepsilon^n$, to be used to find the stress at a strain near to that at initial yield. (It should be noted that in determining σ_1 from σ_{AB}, ε_{AB} and n using equation (6.9) it has only been necessary to assume that equation (4.14) held over a small strain range as ε_{AB} was of the order of 1, the strain corresponding to σ_1.) A curve of stress found in this way for a strain of 0.01 is given in Fig. 6.9 and can be seen to be of the same general shape as the σ_1 curve but much nearer in actual stress values to the lower yield stress curve. The differences in the two curves in Fig. 6.9 could reflect real differences in materials but might also be explained by limitations in the methods used in calculating stresses. In this connection the linear extrapolation used in Fig. 6.8 (broken lines) should be treated with some caution as results for wider ranges of strain-rate, e.g. those in Fig. 6.10 obtained by Campbell and Ferguson (1970) from dynamic shear tests on EN 38 mild steel, show that at higher strain-rates the rate of increase of flow stress with the logarithm of the strain-rate increases markedly. (It should be noted that the results in Fig. 4.7 also show this effect.) This would mean that the values of stress obtained from the extrapolated results would be underestimates. Also, the method of

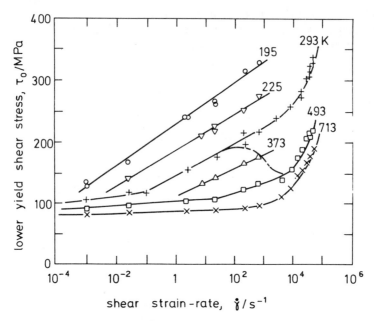

Fig. 6.10 — Experimental lower yield shear stress results: chain dotted line represents stress values corresponding to values of T_{AB} and $\dot{\varepsilon}_{AB}$ in Fig. 6.7.

calculating T_{AB} in which η was taken as unity is open to question as this assumes that all the plastic work of chip formation has occurred at AB. In this way T_{AB} would have been overestimated and hence the stresses found from Fig. 6.8 would be underestimated. Ferguson in written discussion to the paper by Stevenson and Oxley (1970–1971) used the results in Fig. 6.10 to construct the curve (shown chain dotted) corresponding to the values of T_{AB} and $\dot{\varepsilon}_{AB}$ in Fig. 6.7. It can be seen that this curve (Fig. 6.10) again shows a dip and subsequent rise in stress similar to that observed in the σ_1 curve.

In the light of the above results the following qualitative explanation for the influence of cutting speed on the σ_1 curve in Fig. 6.3 can now be given. At low cutting speeds the increase in stress caused by the increase in strain-rate resulting from an increase in cutting speed predominates over the decrease in stress caused by the corresponding rise in temperature (thermal softening). This is because the logarithm of the strain-rate, with which stress increases linearly (Fig. 6.8), changes rapidly with increase in cutting speed at low cutting speeds. For the same increase in cutting speed at higher cutting speeds, the increase in the logarithm of the strain-rate is less and consequently the rise in stress resulting from the increase in strain-rate is reduced. At a certain stage the decrease in stress caused by thermal softening can become of the same order, or even greater, than the increase in stress caused by the increase in strain-rate. For the results in Fig. 6.3 thermal softening clearly predominates over strain-rate hardening at a strain-rate $\dot{\varepsilon}_{AB} \approx 2.0 \times 10^3$/s which corresponds to a cutting speed of about 23 m/min and it is here that the lowest point of the dip in the σ_1 curve occurs. As cutting speed is increased further, T_{AB} (Fig. 6.7) becomes approximately constant while $\dot{\varepsilon}_{AB}$ continues to rise. The strain-rate effect therefore takes over once again and the σ_1 curve starts to rise. It is also at about the point where the temperature becomes approximately constant that n (Fig. 6.3) starts to decrease rapidly and it therefore appears that this is more a strain-rate than a temperature effect.

Further results showing the influence of strain-rate and temperature on flow stress were obtained by Stevenson and Oxley (1970–1971) from their experimental machining results (Figs. 6.1 and 6.2) by considering the flow at the tool–chip interface. To do this they assumed that a plastic state of stress existed in the chip material adjacent to the tool–chip interface over the full contact length with the interface a direction of maximum shear stress and maximum shear strain-rate and, following Boothroyd (1963), that the plastic zone at the interface was rectangular with no sliding at the interface. From these assumptions it follows that the shear flow stress k_{int} and maximum shear strain-rate at the interface are given by

$$k_{int} = \frac{F}{hw} \qquad (6.23)$$

and

$$\dot{\gamma}_{int} = \frac{V}{\delta t_2} \qquad (6.24)$$

and that the values of interface temperature T_{int} can be calculated from equations (6.14) to (6.16). Stevenson and Oxley calculated k_{int}, $\dot{\gamma}_{int}$ and T_{int} (Fig. 6.7) from these equations using the smoothed experimental results in Figs. 6.1 and 6.2 including the experimental values of tool–chip contact length h (Fig. 6.2) which were measured from the length of the wear scar on the tool cutting face. In the calculations δ, the ratio of the thickness of the interface plastic zone to the chip thickness t_2, was taken to be 0.125. The values of F, t_2 and V which were needed in determining k_{int}, etc., were found from the equations

$$F = F_C \sin \alpha + F_T \cos \alpha \qquad (6.25)$$

and

$$t_2 = \frac{t_1 \cos(\phi - \alpha)}{\sin \phi} \qquad (6.26)$$

and equations (2.3) using experimental values of ϕ, F_C and F_T together with the given cutting conditions. In calculating T_{int} an iterative procedure was required as S and K were temperature dependent with in this case the temperature T determining

S and K taken as the mean chip temperature given by $T_C = T_W + \Delta T_{SZ} + \Delta T_C$. The temperature factor ψ in equation (6.14) was taken as unity.

Values of flow stress σ_{int} ($\sigma_{int} = \sqrt{3}k_{int}$) found by Stevenson and Oxley in this way together with values of σ_1 taken from Fig. 6.3 are given in Fig. 6.11. To compare the

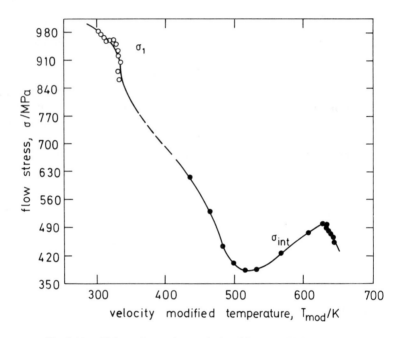

Fig. 6.11 — Values of σ_1 and σ_{int} calculated from machining test results.

two sets of results, which from Fig. 6.7 can be seen to be for different temperature ranges, and to combine the effects of strain-rate and temperature in a single parameter the results have been plotted against velocity-modified temperature. This parameter was first introduced by MacGregor and Fisher (1946) and as mentioned in section 4.5 was used by Fenton and Oxley (1968–1969) in their machining analysis in considering the variations in shear stress at the tool–chip interface with strain-rate and temperature. As used in machining the velocity-modified temperature T_{mod} can be expressed as

$$T_{mod} = T[1 - v\,lg(\dot{\varepsilon}/\dot{\varepsilon}_0)] \qquad (6.27)$$

where $T(K)$ and $\dot{\varepsilon}\,(s^{-1})$ are the testing temperature and strain-rate and v and $\dot{\varepsilon}_0$ are constants for a given material and range of testing conditions. It is assumed that for a given strain the flow stress for a particular material will be a unique function of T_{mod} defined in this way. In calculating T_{mod} for the machining results $\dot{\varepsilon}_0$ was taken as 1/s and v was selected so that two points of equal σ_1 in Fig. 6.3 had the same T_{mod} value which gave v equal to 0.09. For the conditions on AB, T in equation (6.27) was taken as $T_{AB} + 273°C$ with $\dot{\varepsilon}$ equal to $\dot{\varepsilon}_{AB}$ while for the tool–chip interface the values used were $T_{int} + 273°C$ and $\dot{\varepsilon}_{int}$ ($\dot{\varepsilon}_{int} = \dot{\gamma}_{int}/\sqrt{3}$).

The two sets of results (Fig. 6.11) can be seen to fit together quite well, following one curve over the entire range of conditions. This is at first somewhat surprising because the strains at the interface will be significantly greater than one, the strain corresponding to σ_1. The average shear strain at the interface is in fact given approximately by the ratio $h/\delta t_2$ which noting that $h \approx t_2$ gives a shear strain of about 8 if, as assumed by Stevenson and Oxley, δ is equal to 0.125. In reality the strain immediately adjacent to the interface can be many times greater than this as shown by photomicrographs of chip sections and printed grid experimental flow fields. The apparent lack of influence of strain on the σ_{int} values in Fig. 6.11 is, however, consistent with results obtained in materials testing in general which show that, above strains of the order of one, strain has little influence on flow stress. For most metals this is true for all testing conditions; it is particularly true for the high temperatures and strain-rates encountered at the interface.

The results in Fig. 6.11 show a decrease in stress with increase in velocity-modified temperature, as would be expected, up to a modified temperature of about 520 K but above this value, and up to a modified temperature of just over 600 K, the stress increases with increase in modified temperature. This effect is well known from the results of both high and low speed tension and compression tests on mild steels — see, for example, Manjoine (1944), Ohmori and Yoshinaga (1966) and Tanaka and Kinoshita (1967) — and is attributed to dynamic strain ageing or as it is sometimes termed blue-brittleness. Manjoine's results show that as strain-rate is increased the temperature at which dynamic strain ageing occurs is also increased. Noting this Stevenson and Oxley (1970–1971) showed that the interface strain-rates and temperatures at which dynamic strain ageing was observed to occur in machining were consistent with those observed by other workers for dynamic strain ageing conditions in high speed tension and compression tests. Stevenson and Oxley (1973) later demonstrated how dynamic strain ageing conditions could also be achieved in the chip formation zone by pre-heating the work to a sufficiently high temperature.

It can be concluded from the work described so far in this chapter that a carefully designed machining test can provide an effective method for measuring a material's flow stress properties. In fact for the high strains (1 and over), strain-rates (10^3 to 10^6/s) and temperatures (200 to 1000°C) encountered in machining it is difficult to conceive of a more suitable testing method. The philosophy of obtaining a work material's flow stress properties from relatively few machining test results and then applying these to make predictions of cutting forces etc. over a much wider range of conditions as proposed, for example, by Fenton and Oxley (1969–1970), therefore appears a sound one. However, the approach of using machining results to predict machining results clearly invites accusations of tail chasing. It would be far better from the viewpoint of attempting to verify a predictive machining theory if this could be done using flow stress properties obtained from an independent test. The next section considers some of the few such data available which are suitable for this purpose.

6.5 FLOW STRESS DATA DETERMINED FROM HIGH SPEED COMPRESSION TEST RESULTS

Oyane *et al.* (1967) and Oyane (1973) have obtained flow stress data from high speed compression tests for a range of plain carbon steels which are suitable for use in making machining predictions. This is achieved by obtaining values of σ_1 and n, which through equation (4.14) are taken as defining the stress-strain curve of a material for a given strain-rate and temperature, from the compression test results and then expressing these as functions of velocity-modified temperature. Machining predictions were first made in this way by Hastings *et al.* (1974) and Oxley and Hastings (1976). The method of determining σ_1 and n as functions of velocity modified temperature is now described.

In the compression tests the strain-rate (≈ 450/s) was lower than that usually encountered in machining (10^3 to 10^6/s) but the testing temperatures covered a wide range (0 to 1100°C), which makes it possible to extrapolate the compression test results into the machining range by using the velocity-modified temperature parameter given in equation (6.27). This parameter has been used many times before to represent the opposing effects of strain-rate and temperature but normally in situations where temperature is the main variable and there are relatively small variations in strain-rate with the strain-rates not very high. In machining, strain-rates are very high and both temperature and strain-rate vary over large ranges. The compression test results cannot be used to test the validity of equation (6.27) for such conditions because the strain-rate in these tests was constant. However, the results of Campbell and Ferguson (1970) given in Fig. 6.10 which show the effects of very wide ranges of strain-rate and temperature on the lower yield point of a 0.12% plain carbon steel are suitable for this purpose. Lower yield stress points taken from the curves in Fig. 6.10 at strain-rates of 1, 10, 10^2, 10^3, 10^4 and 3.2×10^4/s are replotted against T_{mod}, with as before $v = 0.09$ and $\dot{\varepsilon}_0 = 1$/s, in Fig. 6.12 and can be seen to fit a single curve exceptionally well, thus giving support to the use of equation (6.27) for conditions where strain-rate and temperature vary over large ranges and are of the same order of magnitude as in machining.

Oyane *et al.* (1967) and Oyane (1973) have given results for plain carbon steels with carbon contents in the range 0.16 to 0.55% and their results for a 0.16% carbon steel, plotted on a log–log basis, are given in Fig. 6.13. In the strain range from 0.2 to 0.4 straight lines of best fit have been drawn to approximate the curves for each temperature. That these lines are a good approximation (over a restricted strain range) supports to some extent the use of the linear logarithmic stress against strain relation of equation (4.14) in machining analyses. (Normally the use of this equation has been justified on the basis of slow speed (low strain-rate) compression test results.) A good fit is not obtained over the much wider strain range of the compression tests, chiefly because of the near adiabatic conditions in the tests which cause variations in flow stress with the resulting temperature rise in addition to those caused by strain-hardening. In deriving the constants σ_1 and n for each test it was therefore decided to use the straight lines drawn for the 0.2–0.4 strain interval. The value of n was found directly from the slope and σ_1 by extrapolating the line to a strain $\varepsilon = 1$. In calculating the corresponding T_{mod} values the constants v and $\dot{\varepsilon}_0$ were again taken as 0.09 and 1/s. The strain-rate $\dot{\varepsilon}$ was constant at 450/s and the temperature T (K) was taken as the test starting temperature plus a calculated

Flow stress data

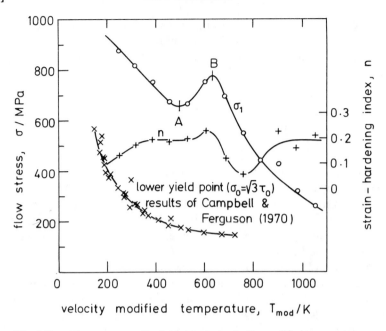

Fig. 6.12 — Flow stress results plotted against velocity-modified temperature.

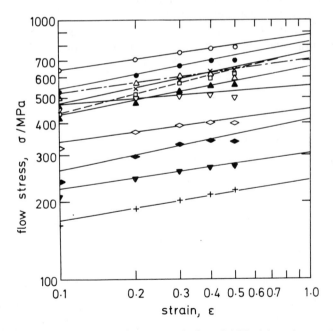

Fig. 6.13 — High speed compression test results for a 0.16% plain carbon steel: ○, 0°C; ●, 100°C; ×, 200°C; ▲, 400°C; □, 500°C; △, 600°C; ▽, 700°C; ◇, 800°C; ◆, 900°C; ▼, 1000°C; +, 1100°C.

temperature rise appropriate to the mid-point of the 0.2–0.4 strain interval. The temperature rise was calculated by assuming that all of the plastic work was converted to heat and that the conditions were adiabatic. The rise ΔT (K) was then given by

$$\Delta T = \frac{1}{\rho S} \int_0^{0.3} \sigma \, d\varepsilon \qquad (6.28)$$

In determining ΔT the density ρ was taken as 7862 kg/m³ and S was calculated from equation (6.18) with T (°C) in this equation taken as the test starting temperature.

The results for σ_1 and n for the 0.16% carbon steel are plotted against T_{mod} in Fig. 6.12 and can be seen to show a clear dynamic strain ageing (blue-brittle) region, where flow stress increases with increase in temperature. This effect is not evident in the results of Campbell and Ferguson (1970), which are also given in Fig. 6.12, because of the very low strain ($\varepsilon < 0.01$) associated with these results. The curves for σ_1 and n have been represented mathematically using sections of polynomials and these have been used with rescaling functions to represent the σ_1 and n curves for the other carbon steels. The rescaling functions, which give an increase in σ_1 and a decrease in n with increase in carbon content, were chosen to give a good fit with the experimental values of σ_1 and n for the 0.35, 0.45 and 0.55% carbon steels derived from the compression test results in the same way as described above. In this way continuous changes in σ_1 and n over the ranges of velocity-modified temperature and carbon content considered were represented by a relatively simple set of functions. The method of obtaining these functions is described in Appendix A3.

7

Predictive machining theory based on a chip formation model derived from analyses of experimental flow fields

7.1 BASIC THEORY AND CALCULATION PROCEDURE

The theory is based on a model of chip formation derived from the slipline field analyses of experimental flow fields described in Chapter 3 and from the strain-rate analysis of experimental flow fields described in Chapter 5. The model used in the analysis is given in Fig. 7.1. Plane strain, steady-state conditions are again assumed to

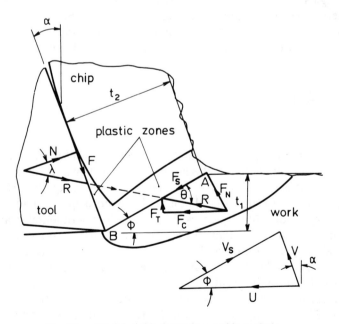

Fig. 7.1 — Model of chip formation used in analysis.

apply and the tool is taken to be perfectly sharp. The plane AB near the centre of the chip formation zone is found from the same construction as used in Fig. 5.2, i.e. in the same way as for the shear plane in Fig. 2.1. AB found in this way can be looked upon as a straight slipline near the centre of the slipline fields for the chip formation zone given in Figs. 3.4 and 3.7. The basis of the theory is to analyse the stress distributions along AB and the tool–chip interface, which is also assumed to be a direction of maximum shear stress and maximum shear strain-rate, in terms of the shear angle ϕ (angle made by AB with cutting velocity), work material properties, etc., and then to select ϕ so that the resultant forces transmitted by AB and the interface are in equilibrium. Once ϕ is known then the chip thickness t_2 and the various components of force can be determined from the same geometric relations as for the shear plane model, that is

$$
\begin{aligned}
t_2 &= t_1 \cos(\phi - \alpha)/\sin \phi \\
F_C &= R \cos(\lambda - \alpha) \\
F_T &= R \sin(\lambda - \alpha) \\
F &= R \sin \lambda \\
N &= R \cos \lambda \\
R &= \frac{F_S}{\cos \theta} = \frac{k_{AB} t_1 w}{\sin \phi \cos \theta}
\end{aligned}
\qquad (7.1)
$$

where all angles, forces, etc., are as shown in Fig. 7.1 and as previously defined.

The method used to analyse the stresses along AB (Fig. 7.1) is essentially the same as the method used by Stevenson and Oxley (1970–1971) which is described in section 6.1. The difference is that the strain-rate along AB is assumed to be given by

$$
\dot{\gamma}_{AB} = C \frac{V_S}{l} \qquad (7.2)
$$

and not by $\dot{\gamma}_{AB} = C' V_S/t_1$ as in equation (6.1). The reason for this change is that the resulting theory is found to predict ϕ more accurately than if the theory was based on equation (6.1). It is interesting to note that in proposing equation (6.1) Stevenson and Oxley (1969–1970) suggested that it might be more appropriate to relate $\dot{\gamma}_{AB}$ to V_S/l than to V_S/t_1 but could not check this with their experimental strain-rate results as ϕ was more or less constant in their tests and therefore l was approximately proportional to t_1. Stevenson and Oxley's strain-rate results (Fig. 5.7) are shown re-plotted against V_S/l in Fig. 7.2 and can be seen again to fit a single straight line reasonably well. From the results in Fig. 7.2, C is found to be approximately 5.9. If equation (7.2) is taken as giving the strain-rate along AB then substituting this equation together with equations (6.2), (6.5) and (6.6) in equation (6.4), in the way described in section 6.1, yields the relation

Sec. 7.1] **Basic theory and calculation procedure** 99

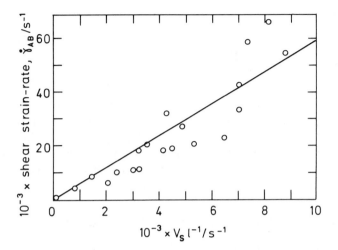

Fig. 7.2 — Experimental strain-rate results.

$$dk/ds_2 = 2Cnk_{AB}/l \tag{7.3}$$

where it will be recalled n is the strain-hardening index in the empirical stress-strain relation $\sigma = \sigma_1 \varepsilon^n$ used to represent the work material flow stress properties, and substituting this for $\Delta k/\Delta s_2$ in equation (4.7) gives

$$\tan \theta = 1 + 2\left(\frac{\pi}{4} - \theta\right) - Cn \tag{7.4}$$

From the geometry of Fig. 7.1 the angle θ can, in the same way as for the shear plane model, also be expressed in terms of other angles by the equation

$$\theta = \phi + \lambda - \alpha \tag{7.5}$$

The temperature at AB which is needed, together with the strain-rate at AB (given by equation (7.2)) and the strain at AB given as before by the equation

$$\gamma_{AB} = \frac{1}{2} \frac{\cos \alpha}{\sin \phi \cos(\phi - \alpha)} \tag{7.6}$$

to determine k_{AB} and n from the work material property curves is found as described in section 6.2 from the equations

$$T_{AB} = T_W + \eta \, \Delta T_{SZ}$$

with (7.7)

$$\Delta T_{SZ} = \frac{1-\beta}{\rho S t_1 w} \frac{F_S \cos \alpha}{\cos(\phi - \alpha)},$$

where T_W is the initial work temperature, F_S is the shear force along AB, η ($0 < \eta \leq 1$) is a factor which allows for the fact that not all of the plastic work of chip formation has occurred at AB, ρ and S are the density and specific heat of the work material and β is the proportion of heat conducted into the work. β is found as before (see section 6.2) from the following empirical equations based on a compilation of experimental results made by Boothroyd (1963):

$$\beta = 0.5 - 0.35 \, lg(R_T \tan \phi) \text{ for } 0.04 \leq R_T \tan \phi \leq 10.0 \qquad (7.8)$$
$$\beta = 0.3 - 0.15 \, lg(R_T \tan \phi) \text{ for } \qquad R_T \tan \phi > 10.0$$

with the non-dimensional thermal number R_T given by

$$R_T = \rho S U t_1 / K \qquad (7.9)$$

where K is the thermal conductivity of the work material. The limits $0 \leq \beta \leq 1$, are also imposed.

In considering the tool–chip interface it is assumed that a plastic state of stress exists in the chip over the full contact length and that the deformation in the chip can be represented by a rectangular plastic zone with no sliding at the interface. It will be noted that this is the model assumed by Stevenson and Oxley (1970–1971) in investigating the influence of strain-rate and temperature on the flow stress at the interface (see section 6.4). There is little information available on the stress distributions at the tool–chip interface for practical cutting conditions. In measuring these experimentally, one technique (Usui and Takeyama 1960, Chandrasekaran and Kapoor 1965) has been to use a tool made of photoelastic material to measure the stresses when cutting a soft work material such as lead at very low cutting speeds. The results obtained in this way show that the normal stress is a maximum near to B (Fig. 7.1) and then reduces more or less linearly over the contact length. By using a special composite (split) tool, Kato et al. (1972) were able to measure stress distributions for practical cutting conditions and found that depending on the work material (steel was not considered) the normal stress distributions varied from approximately triangular, thus agreeing with the photoelastic results, to a combination of approximately uniform for the first half of the contact, reducing linearly over the remainder.† The results from the slow speed machining slipline field of Roth and Oxley (1972) for a free machining steel (Fig. 3.8) show that although the maximum normal stress occurs near to B it reduces much less over the plastic contact

---

† It has recently come to the author's attention that similar results to those of Kato et al. have been obtained for steel work materials using practical cutting conditions by Bagchi and Wright (1987). In this investigation the tool–chip interface stresses have been measured by making use of the photoelastic properties of sapphire cutting tools.

Sec. 7.1] **Basic theory and calculation procedure** 101

length than would be given by a triangular distribution. For a plain carbon steel work material Roth (1969) did in fact find that the normal stress was very nearly constant over this length, reducing to zero rapidly over the short elastic contact length. For simplicity it is assumed in the present analysis that a uniform state of stress exists along the interface. As more information becomes available on interface stresses then it should, if necessary, be relatively easy to modify the theory accordingly. The average temperature at the tool–chip interface, from which the average shear flow stress at the interface is determined, is again taken as

$$T_{int} = T_W + \Delta T_{SZ} + \psi \Delta T_M \qquad (7.10)$$

where ΔT_M is the maximum temperature rise in the chip and the factor ψ $(0 < \psi \leq 1)$ allows for possible variations of temperature along the interface. If the thickness of the tool–chip interface plastic zone is taken as δt_2 and the contact length as h then, as shown in section 6.2, ΔT_M can be calculated from the equation

$$lg\left(\frac{\Delta T_M}{\Delta T_C}\right) = 0.06 - 0.195\delta\left(\frac{R_T t_2}{h}\right)^{1/2} + 0.5 \, lg\left(\frac{R_T t_2}{h}\right) \qquad (7.11)$$

where ΔT_C, the average temperature rise in the chip, is given by

$$\Delta T_C = F \sin \phi / \rho S t_1 w \cos(\phi - \alpha) \qquad (7.12)$$

The tool–chip contact length h is determined from the equation

$$h = \frac{t_1 \sin \theta}{\cos \lambda \sin \phi} \left\{1 + \frac{Cn}{3[1 + 2(\frac{1}{4}\pi - \phi) - Cn]}\right\} \qquad (7.13)$$

which is derived by taking moments about B of the normal stresses on AB to find the position of the resultant force R and noting that for the assumed uniform distribution of normal stress at the interface R intersects the tool face a distance $h/2$ from B. The maximum shear strain-rate at the tool–chip interface, which is also needed in determining the shear flow stress at the interface, is assumed to be given by

$$\dot{\gamma}_{int} = \frac{V}{\delta t_2} \qquad (7.14)$$

where V (Fig. 7.1) is the rigid chip velocity. This equation follows from the assumption that the sliding velocity at the cutting face is zero or in other words that seizure has occurred in the tool–chip contact region. This is consistent with the findings of Trent (1977) who has shown, using optical and electron microscopy to examine chip and tool sections, that the contacting surfaces are, for most practical machining conditions, interlocked with adhering metal penetrating all irregularities in the tool surface. However, for steady-state conditions as assumed in the present analysis material must leave the tool–chip interface plastic zone with a velocity consistent with the rigid body motion of the chip and the sliding velocity cannot be zero over the full contact region. It can, however, be much smaller than the chip velocity over much of this region as can be shown by using a slipline field similar to that suggested by Roth and Oxley (1972) (see Chapter 3). With fields of this kind the sliding velocity increases in moving along the cutting face away from the cutting edge B (Fig. 7.1) and can have very low values, approaching zero, while the velocities at the plastic zone boundary are still consistent with the rigid body motion of a rotating (curled) chip. The associated flow shows similar features to those resulting from seizure with, in particular, the layer of chip material in contact with the tool greatly swept back (retarded) relative to the rest of the chip. Therefore, although equation (7.14) will overestimate $\dot{\gamma}_{int}$ and the cutting face will not be exactly a direction of maximum shear strain-rate, and hence maximum shear stress, because there is a direct strain-rate in this direction, the differences will usually be small and can for the purposes of the analysis be neglected. Further consideration is given to the actual nature of tool–chip interface friction in section 7.3.

Given the flow stress and thermal properties of the work material together with the values of the constants C and δ which determine the strain-rates in the plastic zones the above equations can be used to calculate the shear angle ϕ. To find a solution for a given set of cutting conditions the method used is to calculate, for a range of values of ϕ, the resolved shear stress at the tool–chip interface from the resultant cutting force obtained from the stresses on AB and then for the same range to calculate the temperature and strain-rate at the interface and hence the corresponding values of shear flow stress. The solution is taken as the value of ϕ which gives a shear flow stress in the chip material at the interface equal to the resolved shear stress, as the assumed model of chip formation is then in equilibrium. Once ϕ is known then all other parameters can be determined. The method is now described in detail.

The given information will be the tool rake angle α, the cutting speed U, the thickness t_1 and width w of the undeformed chip, together with the thermal and flow stress properties of the work material and the initial temperature of the work T_W. The constants C in equations (7.2) and (7.4) and δ in equations (7.11) and (7.14) must also be known. For an assumed value of ϕ equation (7.2) can be used to calculate $\dot{\gamma}_{AB}$ with $l = t_1/\sin\phi$ and, from equations (2.3), $V_S = U\cos(\phi - \alpha)$. Equation (7.6) gives the corresponding value of shear strain γ_{AB}. To find the values of σ_1 and n needed in the calculations, which for plain carbon steels are represented by the equations given in Appendix A3, it is necessary to know the value of the velocity-modified temperature

$$T_{mod} = T[1 - v\, lg(\dot{\varepsilon}/\dot{\varepsilon}_0)] \qquad (7.15)$$

where $v = 0.09$ and $\dot{\varepsilon}_0 = 1/s$, for the strain-rate and temperature at AB but at this stage T_{AB} is not known. It is therefore assumed to start that T_{AB} is equal to $T_W(K)$ and this is used together with $\dot{\varepsilon}_{AB}$ ($\dot{\varepsilon}_{AB} = \dot{\gamma}_{AB}\sqrt{3}$) to give the first estimate of T_{mod} at

Sec. 7.1]   **Basic theory and calculation procedure**   103

AB. The values of σ_1 and n corresponding to this T_{mod} are then found and these are used together with ε_{AB} ($\varepsilon_{AB} = \gamma_{AB}/\sqrt{3}$) to find k_{AB} from the relation

$$k_{AB} = \frac{\sigma_1 \varepsilon_{AB}^n}{\sqrt{3}} \qquad (7.16)$$

This value of k_{AB} is then used to find the resultant cutting force R from equations (7.1) with the required value of θ calculated from equation (7.4). The forces needed in the temperature calculations are found from equations (7.1) using this value of R with the angle difference $(\lambda - \alpha)$ found from equation (7.5). Equations (7.7) to (7.9) can now be used to calculate T_{AB} with S and K given by the appropriate equations of the form given in equations (6.18) and (6.21) with T in these equations taken as $T_{AB} = T_W$. The calculations are repeated using this value of T_{AB} to replace T_W as the starting estimate of the temperature at AB and this process is continued until the difference between the starting estimate of T_{AB} and the calculated value differs by less than 0.1 K. The forces, stresses, etc., at this converged temperature value are taken as the appropriate values for the assumed ϕ, and the resolved shear stress at the tool–chip interface τ_{int} is found from the equation

$$\tau_{int} = \frac{F}{hw} \qquad (7.17)$$

with the tool-chip contact length h given by equation (7.13). The temperature and strain-rate at the tool–chip interface are found from equations (7.10) to (7.14) with t_2 and V given by equations (7.1) and (2.3) respectively. An iterative procedure is again necessary in calculating chip temperatures as the thermal properties are temperature dependent. In the method used, the first estimate of the mean chip temperature needed for finding S is taken as equal to $T_W + \Delta T_{SZ}$ with ΔT_{SZ} calculated from equations (7.7) and then equation (7.12) is used to calculate ΔT_C. The process is then repeated using this value of ΔT_C added to $T_W + \Delta T_{SZ}$ as the new estimate and continued until the difference between the estimated and calculated values of $T_W + \Delta T_{SZ} + \Delta T_C$ is less than 0.1 K. Having obtained this value equations (7.10) and (7.11) are used to find T_{int} with K in the expression for R_T taken as that corresponding to a temperature T equal to $T_W + \Delta T_{SZ} + \Delta T_C$. The value of T_{mod} at the tool–chip interface is found by substituting this value of T_{int} (K) together with the corresponding value of $\dot{\varepsilon}_{int}$ ($\dot{\varepsilon}_{int} = \dot{\gamma}_{int}/\sqrt{3}$) into equation (7.15). It is now assumed that the shear flow stress in the chip at the tool–chip interface k_{chip} is given by

$$k_{chip} = \frac{\sigma_1}{\sqrt{3}} \qquad (7.18)$$

where σ_1 corresponds to the value of T_{mod} at the interface. This equation neglects the influence of strain on flow stress above a strain of one (normally $\varepsilon_{int} \gg 1$) which is a reasonable approximation for metals as pointed out in section 6.4 in considering the flow stress results in Fig. 6.11. If strain-hardening does become significant at the interface then as shown by Mathew and Oxley (1981) account can be taken of it in the theory.

Typical curves of τ_{int} and k_{chip} against ϕ obtained in the way described above are given in Fig. 7.3. These are for the 0.16% carbon steel for which the σ_1 and n curves

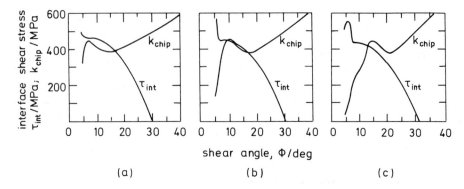

Fig. 7.3 — Curves of shear stress at tool–chip interface showing how solution point values of shear angle are obtained ($\alpha = 5°$; $t_1 = 0.25$ mm; $w = 5.0$ mm): (a) $U = 6$ m/min; (b) $U = 15$ m/min; (c) $U = 60$ m/min.

are given in Fig. 6.12. In the temperature calculations the temperature factors η and ψ were both taken as unity. The solution for ϕ is taken at the intersection of the two curves. Where there is more than one intersection as in Fig. 7.3(b) it might reasonably be expected that the intersection furthest to the right will give the solution as this is the first equilibrium solution reached as ϕ decreases from its relatively high value at the start of a cut. However, as will be seen later there are other possibilities with this type of intersection. It can be seen that depending on the cutting conditions the intersections can occur in the range of k_{chip} values obtained from the part of the σ_1 curve (Fig. 6.12) to the left of A (Fig. 7.3(a)), between A and B (dynamic strain ageing range) (Fig. 7.3(b)) and to the right of B (Fig. 7.3(c)). In this example the change from one type of intersection to the next results from the change in cutting speed, an increase in cutting speed increasing T_{mod} at the tool–chip interface. Changes in rake angle and undeformed chip thickness can act in a similar way by varying T_{mod} at the interface. The significance of changes in cutting conditions changing the type of intersection is considered in section 7.4.

In the first application of the theory Hastings et al. (1974) predicted shear angles and cutting forces for the 0.16% carbon steel for a range of cutting conditions. Their results showed good agreement with experimental results obtained for similar steels. In particular, in agreement with experiment, an increase in cutting speed was predicted to increase the shear angle and decrease the cutting forces so long as the cutting conditions were outside the built-up edge range. In making these predictions it was necessary to know the values of C and δ and Hastings et al. on the basis of the experimental results of Stevenson and Oxley (1969–1970) and Stevenson and Duncan (1973) took these to be 5.9 and 0.05 respectively. Clearly the predictive

Sec. 7.2]   Strain-rate in chip formation zone as part of solution

value of the theory is greatly diminished if machining experiments have first to be made in order to obtain these factors. Fortunately it has been found possible to determine both C and δ as part of the solution and the methods for doing this are described in the next two sections.

7.2 DETERMINATION OF STRAIN-RATE IN CHIP FORMATION ZONE AS PART OF SOLUTION

Oxley and Hastings (1977) noting that the stress boundary condition at B (Fig. 7.1) had not been used in the analysis described in the previous section showed in the following way how this might be used to determine C and hence $\dot{\gamma}_{AB}$ as part of the solution.

The stress equilibrium equation along AB is as pointed out in section 4.2

$$dp = \frac{dk}{ds_2} ds_1 \qquad (7.19)$$

and applying this between A and B and substituting for dk/ds_2 from equation (7.3) gives

$$p_A - p_B = 2Cnk_{AB} \qquad (7.20)$$

The hydrostatic stress p_A can be found from equation (4.6) as before and substituting this in equation (7.20) gives

$$p_B = k_{AB}\left[1 + 2\left(\frac{\pi}{2} - \phi\right) - 2Cn\right] \qquad (7.21)$$

If AB meets the tool–chip interface without changing direction then the normal stress on the tool face at B is given by

$$\sigma'_N = p_B + k_{AB} \sin[2(\phi - \alpha)] \qquad (7.22)$$

where p_B is the hydrostatic stress at B determined from equation (7.21). If on the other hand AB turns through the angle $(\phi - \alpha)$ to meet the interface at right-angles, and this is assumed to occur in negligible distance, then from the appropriate equation of equations (3.1)

$$\sigma'_N = p_B + 2k_{AB}(\phi - \alpha) \qquad (7.23)$$

For many cutting conditions $\phi \approx \alpha$ and equations (7.22) and (7.23) are approximately equal. However, this is not always the case and for consistency with the model used in the theory in which the tool–chip interface is assumed to be a direction of maximum shear stress, equation (7.23) which satisfies this condition will be used. In making calculations it is convenient to combine equations (7.21) and (7.23) to give

$$\frac{\sigma'_N}{k_{AB}} = 1 + \frac{\pi}{2} - 2\alpha - 2Cn \qquad (7.24)$$

For the uniform normal stress assumed along the tool–chip interface the normal stress at B is also given by

$$\sigma_N = \frac{N}{hw} \qquad (7.25)$$

where N (Fig. 7.1) is the normal force at the interface, h is the contact length and w is the width of cut. By imposing the condition $\sigma'_N = \sigma_N$, C can now be determined as part of the solution.

The machining theory extended as above can be applied to predict $\dot{\gamma}_{AB}$ for the conditions used by Stevenson and Oxley (1969–1970) in obtaining their experimental strain-rate results (see Chapter 5) and a comparison made between predicted and experimental results. The method used to find C (and hence $\dot{\gamma}_{AB}$) which satisfies the condition $\sigma'_N = \sigma_N$ for given cutting conditions (α, U, t_1, w and T_W) is to determine the equilibrium solutions (i.e. $\tau_{int} = k_{chip}$) as described in the previous section for a range of values of C and then to calculate the corresponding values of σ'_N and σ_N from equations (7.24) and (7.25). Typical results found in this way are given in Fig. 7.4 and it can be seen that for the conditions considered $\sigma'_N = \sigma_N$ when $C \approx 5.2$.

In their experiments Stevenson and Oxley (1969–1970) used a free machining steel of chemical composition 0.13% C, 1.4% Mn, 0.25% S. 0.019% P as the work material. The flow stress data used with the machining theory for making predictions are for plain carbon steels and for a 0.13% carbon steel the equations used to represent these data (see Appendix A3) show differences in σ_1 and n compared with the values calculated by Stevenson and Oxley (1970–1971) from their machining test results (see section 6.1). The main difference is in the values of n, where for the range of velocity-modified temperatures covered in the machining tests n for the plain carbon steel is approximately double that for the free machining steel. To allow for this the values of n used in making predictions for the free machining steel were therefore taken as half the corresponding values for a 0.13% plain carbon steel found from the equations in Appendix A3. The values for σ_1 for the two materials are for most of the considered conditions in good agreement and in making predictions the values for the plain carbon steel were used without modification. In the calculations δ was taken as 0.02 and the temperature factors η and ψ were both taken as unity.

Predicted and experimental shear angle and strain-rate results are given in Figs. 7.5 and 7.6. The experimental results are those obtained by Stevenson and Oxley (1969–1970). Before considering the strain-rate results it is important to check that

Sec. 7.2] Strain-rate in chip formation zone as part of solution 107

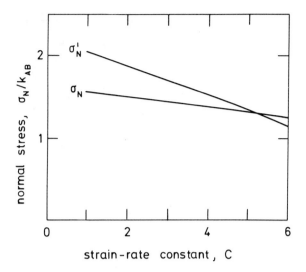

Fig. 7.4 — Values of normal stress showing how solution for strain-rate constant C is obtained: $\alpha = 10°$; $U = 155$ m/min; $t_1 = 0.26$ mm.

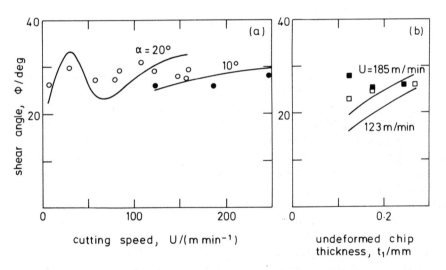

Fig. 7.5 — Predicted and experimental shear angles: (a) $t_1 = 0.26$ mm; $\bigcirc$, $\alpha = 20°$; $\bullet$, $\alpha = 10°$; (b) $\alpha = 10°$; $\square$, $U = 185$ m/min; $\blacksquare$, $U = 123$ m/min.

the machining theory predicts shear angles that agree reasonably well with those measured experimentally by Stevenson and Oxley. It should be noted that the experimental shear angles given in Fig. 7.5 were measured from 'quick-stop' chip

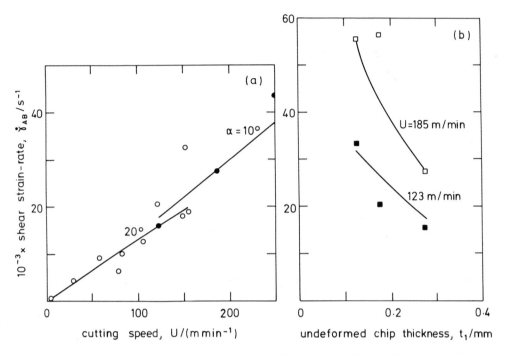

Fig. 7.6 — Predicted and experimental strain-rates: (a) $t_1 = 0.26$ mm; $\bigcirc$, $\alpha = 20°$; $\bullet$, $\alpha = 10°$; (b) $\alpha = 10°$; $\square$, $U = 185$ m/min; $\blacksquare$, $U = 123$ m/min.

sections and are therefore only representative of very small test specimens. Thus it is not surprising that the results show considerable scatter. To overcome this, in calculating material properties from their machining results Stevenson and Oxley (1970–1971) made an additional series of tests to measure average shear angles (from average chip thicknesses) and cutting forces for relatively long cuts, hence effectively giving larger test specimens and averaging out the variations in material properties. These results (see Fig. 6.2) show well-defined trends compared with those in Fig. 7.5 which nevertheless can be seen to be quite near to the predicted values for most of the conditions considered. The results in Fig. 7.6 show that the predicted and experimental strain-rates are at least of the same order of magnitude and if allowance is made for the scatter in the experimental strain-rates for the same reasons as above then it can be concluded that the agreement between the predicted and experimental results is very good. It is worth noting that the corresponding calculated values of C vary from about 3.3 to 7.1 with an average value of just under 4.7 which does not differ greatly from the value $C \approx 5.9$ given by the experimental results in Fig. 7.2.

In view of the success in predicting strain-rates it is of interest to apply the extended theory to investigate the influence on strain-rate, etc., of allowing the work material to approach the ideal constant flow stress material normally assumed in slipline field theory. It has been argued in attempting to justify this assumption that at the high rates of strain encountered in most metal working processes the slope of the corresponding stress–strain curve will decrease ($n \to 0$) and therefore that the variable flow stress terms in equations (3.1) can be neglected. If this is accepted then the slipline fields based on equations (2.1) might be expected to give stress distributions approximating closely to the actual distributions. The dangers in this argument can be seen by considering the construction of the variable flow stress terms which clearly need not have magnitudes directly proportional to the slope of the corresponding stress–strain curve. In applying the theory to this problem it must be emphasised that the chip formation model on which it is based is approximate. In particular the normal stress distribution at the tool–chip interface is assumed and not derived and changes in this will affect the predicted results. This must therefore be borne in mind in considering the following results.

The method used in allowing n to approach zero is to use the equations obtained from Appendix A3 for the 0.13% plain carbon steel but with n multiplied by a decreasing fraction; σ_1 is not adjusted and is therefore still allowed to vary with strain-rate and temperature. Predicted results obtained in this way for the shear strain-rate and change in hydrostatic stress along AB (Fig. 7.1) for a standard set of cutting conditions are given in Fig. 7.7. It can be seen that $\dot{\gamma}_{AB}$ becomes extremely large for $n < 0.01$ while p_A-p_B shows little change even for exceedingly small values of n. It can be concluded that for the chip formation model assumed in the analysis there would be no justification in assuming that the hydrostatic stress became constant along AB simply because n approached zero.

7.3 NATURE OF TOOL–CHIP INTERFACE FRICTION AND MODELLING IT AS A MINIMUM WORK PROCESS

It is now generally accepted that for normal cutting conditions the real and apparent areas of contact are equal over much of the tool–chip contact region with plastic deformation occurring in the adjoining layer of chip. The reasons for the severity of the frictional contact are the chemically clean nature of the newly formed chip contacting surface and tool face, on which any contaminant will be removed early in a cut, and the high normal pressure between chip and tool which is of the order of twice the shear flow stress along the interface (see Fig. 3.8). As the normal pressure drops off towards the end of tool–chip contact then conditions become less severe and, as would be expected, the ratio of real to apparent areas reduces and approaches a value appropriate to normal sliding conditions. In the present analysis this latter region has been ignored on the grounds that it is relatively small and therefore transmits only a small part of the resultant force. It is therefore the nature of the friction in the region where the real and apparent areas of contact are equal that is of main interest. There is no dispute that the layer of chip material in contact with the cutting face over this region is greatly swept back relative to the rest of the chip. However, while Trent (1977) and others, including Zorev (1963) and Wallace and Boothroyd (1964), have taken this as indicative that seizure has occurred at the interface, Enahoro and Oxley (1966) have argued that it only indicates that the flow at the interface is retarded relative to the rest of the flow and that indeed seizure is unacceptable for steady-state conditions.

Evidence to support the idea that for steady-state conditions, with no build up of work material on the cutting face, there must be relative movement at the actual

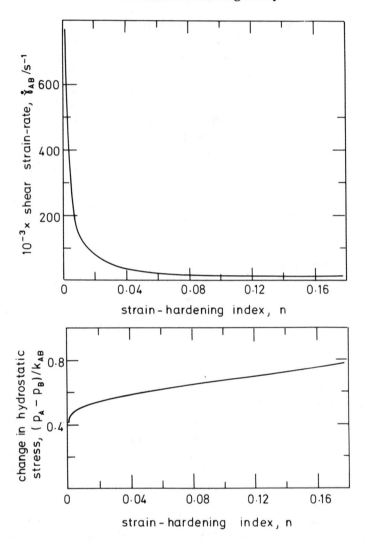

Fig. 7.7 — Predicted results showing the influence of letting the strain-hardening index approach zero.

interface has been provided by Doyle et al. (1979) who have used a transparent sapphire cutting tool to observe the frictional interactions occurring at the interface directly during metal cutting experiments. Their results show that, although in the immediate vicinity of the cutting edge there is intimate contact (real and apparent areas of contact are equal), in agreement with the results of Trent (1977), relative movement is always observed at the interface in this region. To explain this movement for such extreme conditions, Doyle et al. have likened the process to the translation of elastic buckles on the chip surface along the interface in a manner resembling the movement of a ruck in a carpet and in support of this have pointed to the observation of such an effect by Schallamach (1971) when highly elastic rubbers slide against smooth hard surfaces. However, they considered this mechanism doubtful for metals where the elastic strains that can be sustained are smaller by several orders of magnitude than for rubbers. Wright *et al.* (1979), in considering the results of Doyle *et al.*, have noted that plastic deformation occurs in the layer of chip material adjacent to the interface even with relative movement at the interface and have attributed this retardation of the flow to bonding interaction of the chip with the tool. They have proposed, on the basis of this work and the machining results of others, that the contact region at the tool–chip interface consists of a large number of microregions with at any instant a proportion of these exhibiting full seizure; the remainder exhibit interfacial sliding, although it is not clear by what mechanism sliding takes place. Thus, it is argued that for chemically clean surfaces the ratio of seized area to the total contact area approaches unity and there is no sliding at the interface whereas reduction of this ratio by, for example, the presence of contaminants at the interface allows some sliding with, in the limit (perfect lubrication), full sliding. In proposing this model Wright *et al.* are clearly thinking of friction in terms of the classical adhesion model of Bowden and Tabor (1950). In this the frictional force is explained as the force needed to shear the welded junctions which are assumed to be formed by adhesion at the tips of contacting asperities. For normal sliding this seizure condition will be local. However, if the real and apparent areas are equal then seizure will presumably have occurred over the full contact region as proposed by Trent (1977). Indeed, it does not appear that the adhesion model can give other than seized contact regions. Therefore, to sustain the concept of relative movement at the tool–chip interface requires an asperity deformation model which need not involve seizure. Such a model is now considered.

Figure 7.8 gives the steady-state, plane strain (rigid–perfectly plastic) slipline field and the corresponding hodograph suggested by Challen and Oxley (1979) as the basic asperity deformation model for sliding, metallic friction. In this the hard asperity is assumed to be rigid so that the plastic deformation of only one surface need be considered. The field is similar to the fields proposed by Green (1955) for a weak junction, which apparently he made little use of, by Collins (1972) in considering rolling contact and by Johnson and Rowe (1967–1968) to represent the wave that can occur ahead of the die in wire drawing. The interface DE between the hard and soft asperities and the surface EA, which is taken to be stress free, are both assumed to be straight with their directions defined by the angles α and η measured from the sliding direction, which for convenience is represented by the velocity U of the soft material, the hard asperity being assumed to be stationary. The plastic region ABCDE is that existing once sliding is occurring and after the initial — in terms of the adhesion model — junction growth period is over. The angle η will therefore in general be greater than the initial slope of asperities on the soft surface but the angle α will remain unchanged and will be related to the surface roughness of the hard surface. In constructing the slipline field the independent variables are taken to be the angle α and the normalised strength f of the interfacial film along ED, defined as the ratio of the shear strength of the film τ to the shear flow stress of the soft material k, that is $f = \tau/k$, with $0 \leq f \leq 1$. With this model the deformation is represented as a plastic wave and the straight line joining A and D must be parallel to U to satisfy volume constancy. Also, the shear force on the soft asperity at the interface must act

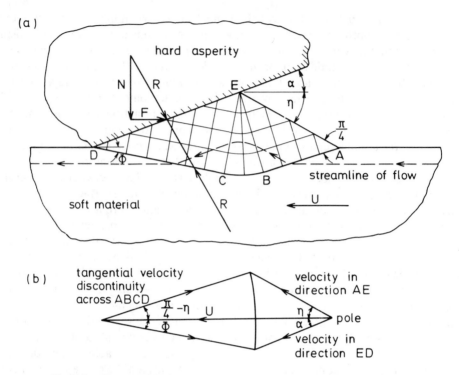

Fig. 7.8 — Wave model of asperity deformation: (a) slipline field; (b) hodograph.

in the direction DE to oppose motion, with the value of f determining the inclination of the sliplines to DE. These conditions, together with the condition that the sliplines must be inclined at an angle of $\pi/4$ to the stress-free surface EA to satisfy equilibrium, define the slipline field and it follows from geometry that

$$\alpha + \Phi = \tfrac{1}{2}\cos^{-1} f \tag{7.26}$$

and

$$\eta = \sin^{-1}[(1-f)^{-1/2}\sin\alpha] \tag{7.27}$$

where Φ is the angle between CD and U and is measured positive as shown in Fig. 7.8. The hodograph shows that there is a tangential discontinuity of velocity across the boundary slipline ABCD so that material that enters and leaves the field with a velocity U flows in the directions AE and ED in the regions ABE and CDE and along a curved path in the centred fan region BCE, as shown by the typical streamline given in Fig. 7.8. The range over which the solution can apply is determined from the condition that Φ must be positive so that material can flow into and out of the wave. Taken in conjunction with equations (7.26) and (7.27) this gives $\alpha \leqslant \eta < \pi/4$. The friction force with this model can be looked upon as the force needed to push plastic waves in the soft material ahead of asperities on the hard surface. Challen et al. (1984) have shown that the theory based on the slipline field gives coefficients of friction and hence frictional forces in good agreement with experimental results.

A particularly interesting property of the slipline field (Fig. 7.8) is that it retards the flow, with material within the deforming region ABCDE travelling at a reduced velocity compared with material outside it. This can be seen from the hodograph which shows that the reduction in velocity depends on the values of α and Φ with an increase in α or a decrease in Φ both increasing the degree of retardation. Even for smooth surfaces ($\alpha \to 0$) the velocity can be reduced to near zero if $f \to 1$ and hence from equation (7.26) $\Phi \to 0$, while continuity of flow is still maintained, and it is only when $\Phi = 0$ that flow ceases and (local) seizure occurs. When the real and apparent areas of contact are equal as for the flow at the tool–chip interface Challen and Oxley (1984a) have shown that a slipline field similar to that in Fig. 7.8 can still be used to represent the flow although in this case material will fill the irregularities (valleys) on the tool face and AE will no longer be stress free. With this field, material can flow in and out of the irregularities so long as Φ is positive and the reduction in velocity will depend on the values of α and f in the same way as for the field in Fig. 7.8. Near-seizure conditions ($\Phi \to 0$) can therefore be accounted for by a steady-state model in which the sliding velocity, although approaching zero, is not zero and it is only when $\Phi = 0$ that true seizure is predicted to occur.

Attention is now turned to considering how the thickness of the plastic zone at the interface δt_2 and hence $\dot{\gamma}_{int}$ might be determined from the analysis. From the machining theory it can be shown that as δ is reduced the calculated values of strain-rate and temperature at the tool–chip interface both increase, with $\dot{\gamma}_{int}$ tending to infinity and T_{int} to some finite value as δ approaches zero. The combined effect of these strain-rate and temperature changes with δ is to give curves of interface T_{mod} of the shape shown in Fig. 7.9 with T_{mod} passing through a maximum at some particular value of δ. The results show that the maximum value of T_{mod} increases with increase in cutting speed U and undeformed chip thickness t_1 while the corresponding values of δ decrease. If attention is limited to conditions where T_{mod} at the interface is above the dynamic strain ageing range then the curves for the shear flow stress k_{chip}, the chip thickness t_2 and the rate of work, both total $F_C U$ and frictional FV, corresponding to a typical T_{mod} curve are as shown in Fig. 7.10. (The results in both Figs. 7.9 and 7.10 are for the 0.16% carbon steel.) The minimum value of k_{chip} must of course coincide with the maximum in T_{mod} and as far as can be judged within the accuracy of the calculations this is also the case for the minimum values of t_2, and $F_C U$ and FV. Oxley and Hastings (1976) have proposed that in practice δ will take up values satisfying this minimum work condition and have shown that predicted values of t_2 and δt_2 obtained in this way agree well with experimental values measured from photomicrographs of chip sections. More recent results confirming this are given in the next section.

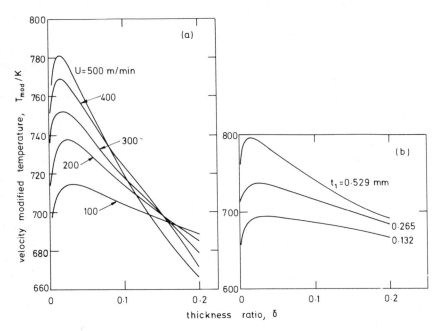

Fig. 7.9 — Curves of velocity-modified temperature: (a) $\alpha = 5°$; $t_1 = 0.265$ mm; $w = 3.0$ mm; (b) $\alpha = 5°$; $U = 200$ m/min; $w = 3.0$ mm.

7.4 COMPARISON OF PREDICTED AND EXPERIMENTAL CHIP GEOMETRY AND CUTTING FORCE RESULTS

A detailed comparison of results predicted from the machining theory with experimental results has been made by Hastings *et al.* (1980). This work is now described. The work materials used in the experiments were plain carbon steels of chemical composition 0.20% C, 0.15% Si, 0.015% S, 0.72% Mn, 0.015% Al and 0.38% C, 0.1% Si, 0.77% Mn, 0.015% P. The curves of σ_1 and n for these materials obtained from the equations in Appendix A3 are given in Fig. 7.11. The thermal properties S and K are represented by equations (6.18), (6.21) and (6.22). The density ρ was taken as 7862 kg/m³ for both materials. In making calculations the temperature factors η and ψ in equations (7.7) and (7.10) were both taken as 0.7 on the basis of the finite element temperature results of Tay *et al.* (1976) as discussed in section 6.2. The complete predictive theory including the determination of C and δ as part of the solution was used to calculate ϕ, cutting forces, etc.

In making calculations for given values of tool rake angle α, cutting speed U, undeformed chip thickness t_1, width of cut w and initial work temperature T_W (taken as 20°C in all of the calculations in this and the following chapters) the following procedure is used. For a given δ the equilibrium ($\tau_{int} = k_{chip}$) values of ϕ are found for a range of values of C and the required value of C is determined from the condition $\sigma'_N = \sigma_N$. This is repeated for different values of δ and the final solution for ϕ, cutting forces, etc., is taken at the value of δ which maximises T_{mod} and thus minimises k_{chip}, etc. A summary of the method of making the calculations is given by the chart in Fig. 7.12.

Sec. 7.4] Comparison of predicted and experimental chip geometry

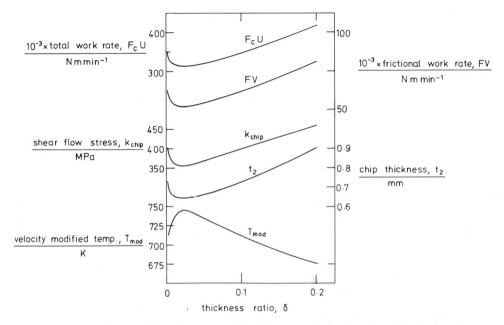

Fig. 7.10 — Curves of shear flow stress, chip thickness and rate of total and frictional work corresponding to a typical velocity-modified temperature curve: $\alpha = 5°$; $U = 250$ m/min; $t_1 = 0.265$ mm; $w = 3.0$ mm.

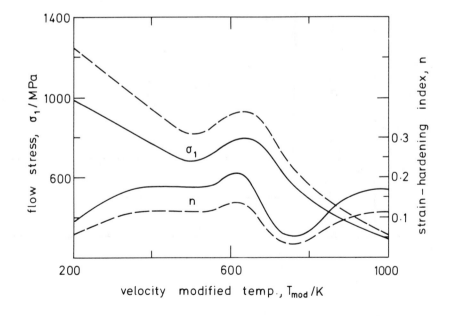

Fig. 7.11 — Flow stress results plotted against velocity-modified temperature: ———, 0.2% carbon steel; – – –, 0.38% carbon steel.

116 Predictive machining theory [Ch. 7

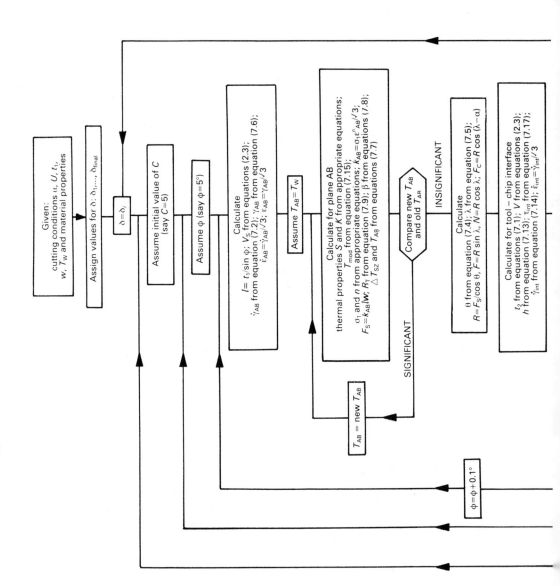

Sec. 7.4] **Comparison of predicted and experimental chip geometry** 117

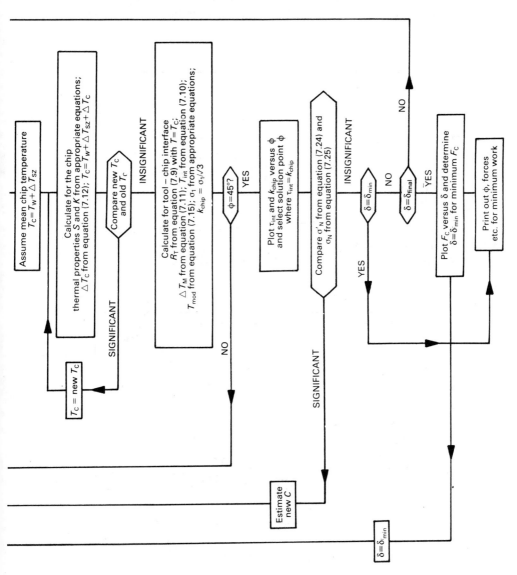

Fig. 7.12 — Summary of method of machining calculations.

A computer program was written for making these calculations and when run on a Cyber 171 mainframe computer 180 s of CPU time was needed on average to obtain results for a given set of conditions. Predicted results, covering the same range of conditions as used in the experiments, for the cutting forces F_C and F_T, chip thickness t_2 and tool–chip interface plastic zone thickness δt_2 are represented by the full lines in Figs. 7.13, 7.14 and 7.15. The broken lines in Fig. 7.15, which lie above

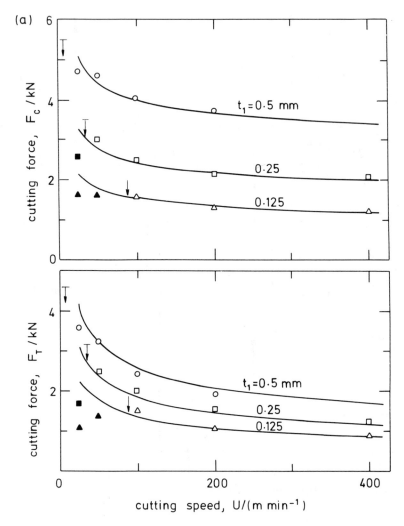

Fig. 7.13(a) — Predicted and experimental cutting forces. Arrows indicate cutting speed at which built-up edge is predicted to disappear: ⬇, $T_{mod} = 700$ K; ↓, $T_{int} = 1000$ K. ●, ■, ▲, experimental results for which a built-up edge occurred ($w = 4.0$ mm): 0.2% carbon steel; $\alpha = -5°$.

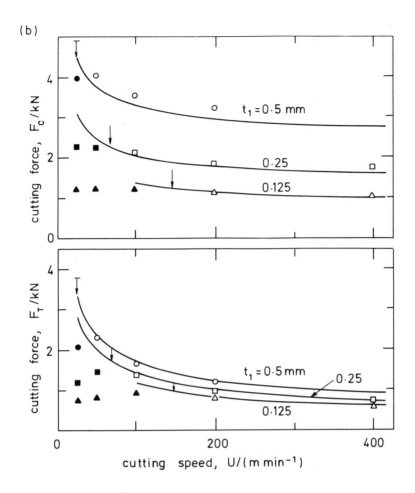

Fig. 7.13(b) — Predicted and experimental cutting forces. Arrows indicate cutting speed at which built-up edge is predicted to disappear: ⊤, $T_{mod} = 700$ K; ↓, $T_{int} = 1000$ K. ●, ■, ▲, experimental results for which a built-up edge occurred ($w = 4.0$ mm): 0.2% carbon steel; $\alpha = 5°$.

and below the full line representing the minimum work results, give δt_2 values which correspond to a 1% deviation from the minimum work points on the total rate of work ($F_C U$) curves. The corresponding variations in F_C, F_T and t_2 are too small to be shown in Figs. 7.13 and 7.14.

The experiments were made on a Heindenreich and Harbech, Gildemeister M530 lathe. This was integrated with a Seimens type IGF9 194 variable speed motor, rating up to 37 kW with a rotational speed of 0–5600 rev/min. A digital display panel for the rotational speed was provided and control of the cutting velocity (taken as equal to the mean circumference of the cut multiplied by the number of the

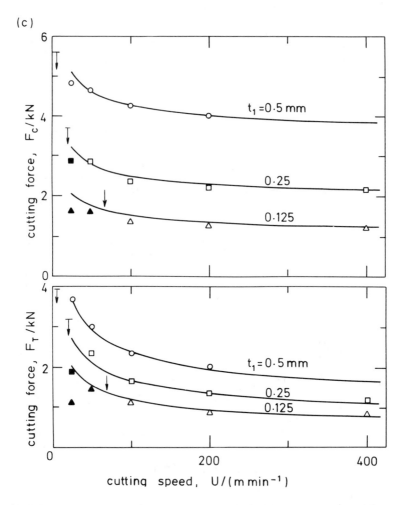

Fig. 7.13(c) — Predicted and experimental cutting forces. Arrows indicate cutting speed at which built-up edge is predicted to disappear: ⊤, $T_{mod} = 700$ K; ↓, $T_{int} = 1000$ K. ●, ■, ▲, experimental results for which a built-up edge occurred ($w = 4.0$ mm): 0.38% carbon steel; $\alpha = -5°$.

revolutions per minute of the workpiece†) could thus be achieved more easily. Cutting forces were measured using a Kistler type 9257A three-component piezoelectric dynamometer which was mounted in place of the front toolpost on the saddle of the lathe. The induced signals were processed and amplified by three Kistler type 5001 charge amplifiers each connected to an output channel of the dynamometer. The final results of the force signals were recorded on a Rikadenki multipen chart recorder. A conventional turning process was used with the main cutting edge of the

---

† As defined in section 1.3 the cutting velocity U is the velocity of the work relative to the tool but in turning the feed velocity is normally negligible when compared with the work velocity and can be ignored.

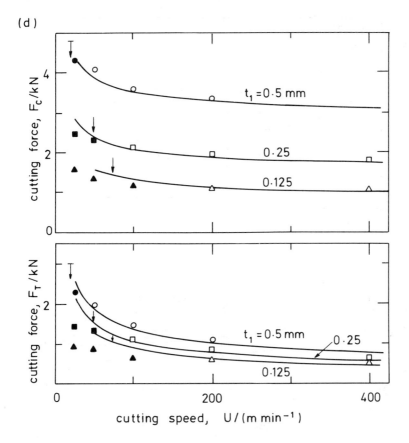

Fig. 7.13(d) — Predicted and experimental cutting forces. Arrows indicate cutting speed at which built-up edge is predicted to disappear: ⟱, $T_{mod} = 700$ K; ↓, $T_{int} = 1000$ K. ●, ■, ▲, experimental results for which a built-up edge occurred ($w = 4.0$ mm): 0.38% carbon steel; $\alpha = 5°$.

tool set normal to the cutting and feed (parallel to the machined bar axis) velocities. The conditions were only approximately orthogonal in as much as a bar, not a tube, was used and therefore there was some cutting at the small secondary cutting edge approximately parallel to the feed direction. However, preliminary tests that were made under the same cutting conditions on a bar and on the same bar undercut in such a way as to provide a short length of tube showed that F_C and F_T were less than 3% greater for the bar than for the tube. The cutting conditions used in the experiments were as follows: $\alpha = 5°$ and $-5°$; t_1 (equal to the axial feed measured in mm/rev) = 0.125, 0.25 and 0.5 mm; $U = 25$, 50, 100, 200 and 400 m/min (for $t_1 = 0.5$ mm the maximum speed was limited to 200 m/min because of power restrictions) and $w = 4.0$ mm. The cutting tools used were tungsten carbide 'throwaway' tips with a 6° clearance angle and negligible nose radius. The duration of each test was short (less than 20 s) as approximately steady-state conditions, judged from cutting forces, were very rapidly achieved.

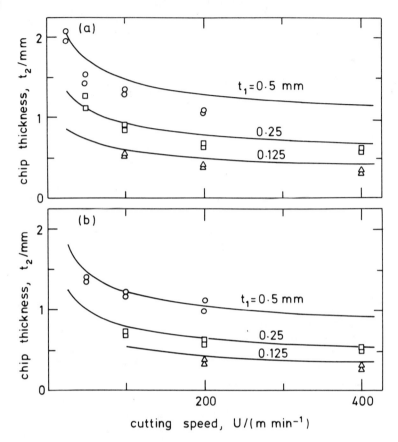

Fig. 7.14 — Predicted and experimental chip thicknesses ($w = 4.0$ mm): (a) 0.2% carbon steel; $\alpha = -5°$; (b) 0.2% carbon steel; $\alpha = 5°$.

The experimental values of F_C and F_T in Fig. 7.13 are the average values for each cutting condition. For conditions in which there was no built-up edge the fluctuation in forces was less than 10%; however, when a built-up edge did occur this increased to as much as 50%. Experimental values of t_2 and δt_2 were measured on a Nikon shadowgraph, equipped with a measuring stage, from chip sections that had been mounted, polished and etched by standard techniques. For each test condition five measurements were made of t_2 and δt_2 at various points along the chip to show possible variation during a test. The method of measurement can be seen from the photomicrograph given in Fig. 7.16(a) of a typical chip section. The chip thickness t_2 was taken as the distance from the back face of the chip to a line drawn to average out the relatively rough outer chip surface. From the photomicrograph it can be seen that in the bulk of the chip the grains exhibit a markedly preferred orientation which results from the deformation in the plastic zone in which the chip is formed. Material that has also passed through the tool–chip interface plastic zone can be identified reasonably clearly by noting where the grains have been swept back relative to the

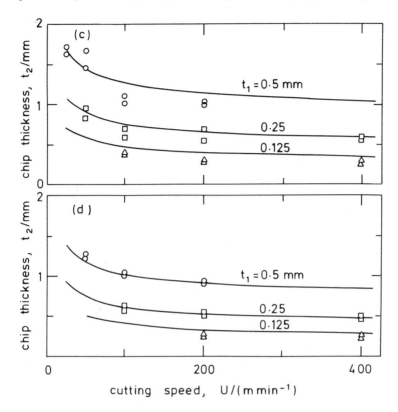

Fig. 7.14 (cont.) — Predicted and experimental chip thicknesses ($w = 4.0$ mm): (c) 0.38% carbon steel; $\alpha = -5°$; (d) 0.38% carbon steel; $\alpha = 5°$.

rest of the microstructure as indicated by the line in Fig. 7.16(a) from which $\delta t_2''$ is measured. Oxley and Hastings (1976) reasoned that $\delta t_2''$ found in this way would represent the maximum thickness of what, in practice, was approximately a triangular plastic zone (Fig. 7.1) and assumed that δt_2, as an average thickness, would be half this value. It can also be seen from Fig. 7.16(a) that a region of intense deformation, of thickness $\delta t_2'$, can be identified and Hastings et al. (1980) decided to measure both this thickness and $\delta t_2''$ for comparison with the predicted values of δt_2. To show the spread of results for the five readings of t_2, $\delta t_2'$ and $\delta t_2''$ the largest and smallest values have been given in Figs. 7.14 and 7.15. The photomicrographs could also be used to observe whether a built-up edge as shown in Fig. 7.17 had been present in a test. To show such results, filled-in symbols in Fig. 7.13 have been used to represent the experimental cutting forces for those conditions in which a built-up edge occurred. Because of the large variations in chip geometry when machining with a built-up edge the corresponding values of t_2, $\delta t_2'$ and $\delta t_2''$ have been omitted from Figs. 7.14 and 7.15.

The agreement between the predicted and experimental cutting forces (Fig. 7.13) can be seen to be exceptionally good for those conditions in which there was no built-

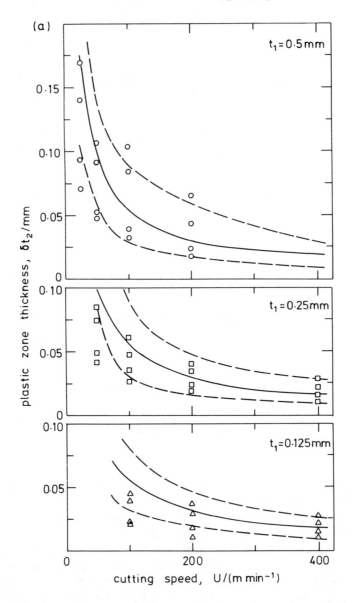

Fig. 7.15(a) — Predicted and experimental tool–chip interface plastic zone thicknesses. ———, minimum work values; – – –, 1% deviation from minimum work values. Top set of experimental points are for $\delta t_2''$ and bottom set are for $\delta t_2'$ ($w = 4.0$ mm): 0.2% carbon steel; $\alpha = -5°$.

up edge, with both F_C and F_T decreasing with increase in tool rake angle α and cutting speed U. The results also show the usually observed size effect in machining with F_C increasing less than in direct proportion to the undeformed chip thickness t_1. If $F_C \propto t_1$ then an increase in t_1 from 0.125 to 0.5 mm would be expected to give a

Sec. 7.4]   Comparison of predicted and experimental chip geometry   125

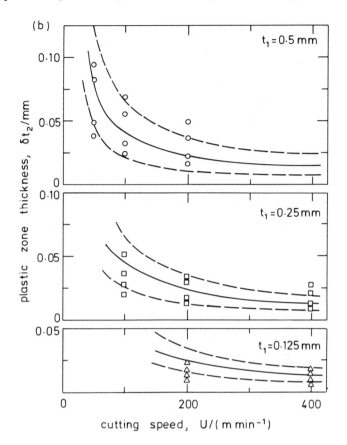

Fig. 7.15(b) — Predicted and experimental tool–chip interface plastic zone thicknesses. ———, minimum work values; — — —, 1% deviation from minimum work values. Top set of experimental points are for $\delta t_2''$ and bottom set are for $\delta t_2'$ ($w = 4.0$ mm): 0.2% carbon steel; $\alpha = 5°$.

four-fold increase in F_C whereas, for example, from Fig. 7.13(a) for $U = 100$ m/min the increase is only (approximately) from 1500 to 4000 N. There is little change in cutting forces with carbon content which is somewhat surprising considering the increase in strength (σ_1 in Fig. 7.11) with increase in carbon content. However, the reason for this can be seen from the results for chip thickness t_2 (Fig. 7.14) which show that t_2 decreases with increase in carbon content, thus decreasing the area of the plane AB (Fig. 7.1) and tending to cancel out the increase in k_{AB}. Similarly, the observed decrease in t_2 with increase in α and U (Fig. 7.14) is consistent with the corresponding changes in F_C and F_T. The agreement between the predicted and experimental values of t_2 is not as good as for forces particularly for $t_1 = 0.5$ mm, $\alpha = -5°$, where the experimental values of t_2 are considerably less than the predicted values. This can be largely explained by the measured increase in width w from 4.0 to approximately 5.0 mm during chip formation for these conditions, thus reducing the experimental values of t_2. It should be noted that for $t_1 = 0.5$ mm the ratio w/t_1 was

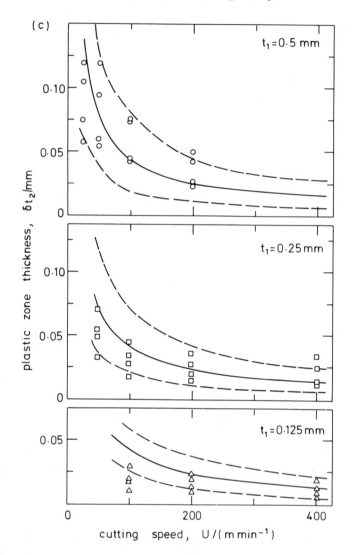

Fig. 7.15(c) — Predicted and experimental tool–chip interface plastic zone thicknesses. ———, minimum work values; – – –, 1% deviation from minimum work values. Top set of experimental points are for $\delta t_2''$ and bottom set are for $\delta t_2'$ ($w = 4.0$ mm): 0.38% carbon steel; $\alpha = -5°$.

only 8 which is small for approximately plane strain conditions and that for $t_1 = 0.125$ and 0.25 mm the measured increase in w was less than 5%. When the large increase in w for $t_1 = 0.5$ mm is taken into consideration it is encouraging from a practical viewpoint that the predicted forces agree so well with the experimental values in this range.

It is difficult, because of the complex interaction between the chip formation and tool–chip interface regions, to give simple physical explanations in terms of the

Sec. 7.4] Comparison of predicted and experimental chip geometry

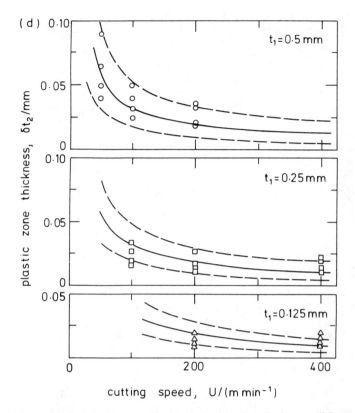

Fig. 7.15(d) — Predicted and experimental tool–chip interface plastic zone thicknesses. ———, minimum work values; – – –, 1% deviation from minimum work values. Top set of experimental points are for $\delta t_2''$ and bottom set are for $\delta t_2'$ ($w = 4.0$ mm): 0.38% carbon steel; $\alpha = 5°$.

theory of the changes in cutting forces and chip thickness with cutting speed, etc. However, it appears that the decrease in t_2 and hence forces with increase in U result mainly from the decrease in k_{chip} caused by the increase in T_{int}, this outweighing the corresponding increase in strain-rate $\dot{\gamma}_{int}$ and thus increasing T_{mod} at the tool–chip interface. (It should be noted that for the cutting conditions considered the values of T_{mod} in the chip formation zone fall to the left of the dynamic strain ageing range in Fig. 7.11 while in the absence of a built-up edge the T_{mod} values at the tool–chip interface mainly fall to the right of this range.) The size effect can also be explained in this way as an increase in t_1 increases T_{mod} at the interface thus decreasing k_{chip} and increasing the chip thickness ratio t_1/t_2 and giving a smaller increase in cutting force than would have resulted had this ratio remained constant. The influence of carbon content on shear angle ϕ and therefore t_2 can be seen from Fig. 7.18 which, for the same cutting conditions, gives the τ_{int} and k_{chip} curves corresponding to the final solution for both the 0.2 and 0.38% carbon steels. The τ_{int} curve is higher for the 0.38% carbon steel as might be expected because of the higher values of k_{AB} while the k_{chip} curve is actually lower because of the higher temperatures at the tool–chip

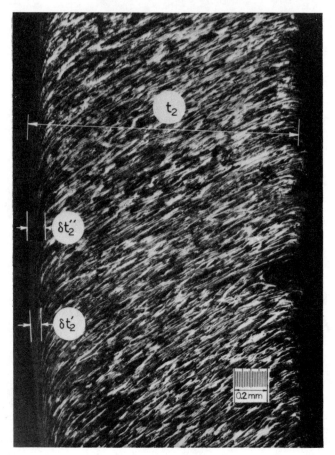

Fig. 7.16(a) — Photomicrograph of chip section (0.38% carbon steel; $\alpha = -5°$; $t_1 = 0.5$ mm); $U = 25$ m/min.

interface. As a consequence of this the intersection of the curves ($\tau_{int} = k_{chip}$) for the 0.38% carbon steel is moved to the right of that for the 0.2% carbon steel and hence gives a larger value of ϕ and a thinner chip.

If in considering the tool–chip interface plastic zone thickness results (Fig. 7.15) it is assumed as by Oxley and Hastings (1976) that $\delta t_2''$ (Fig. 7.16) is the maximum thickness of what is actually a triangular-shaped plastic zone then the $\delta t_2'$ experimental results which can be seen to be approximately half the corresponding values of $\delta t_2''$ would be the more appropriate results for comparison with the predicted values of δt_2. The actual shape of this region is uncertain for the conditions considered and it appears sufficient at this stage to note that all of the experimental results fall within or close to the 1% deviation from minimum work bands and therefore that either $\delta t_2'$ or $\delta t_2''$ is suitable for comparing with the predicted trends in δt_2. It can be seen (Fig. 7.15) that δt_2 is predicted to decrease with increase in U, α and carbon content and with decrease in t_1 and that the experimental results follow all of these trends. The

Sec. 7.4]   **Comparison of predicted and experimental chip geometry**   129

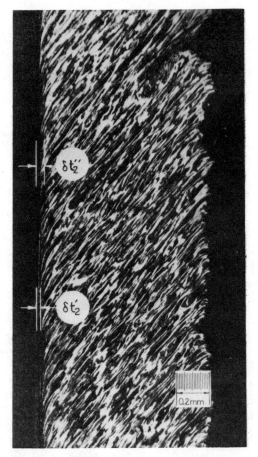

Fig. 7.16(b) — Photomicrograph of chip section (0.38% carbon steel; $\alpha = -5°$; $t_1 = 0.5$ mm): $U = 200$ m/min.

decrease in δt_2 with increase in U is very marked for speeds below 200 m/min and, for example, for $t_1 = 0.5$ mm in Fig. 7.15(c) δt_2 is predicted to change from (approximately) 0.125 to 0.025 mm over the speed range from 25 to 200 m/min. The experimental results show a similar decrease in thickness (the corresponding photomicrographs for $U = 25$ and 200 m/min are given in Figs. 7.16(a,b)), thus confirming one of the main features of the minimum work predictions.

Referring to Fig. 7.11 it can be seen that when the values of T_{mod} at the tool–chip interface are higher than (approximately) 700 K it follows that the layer of chip material adjacent to the interface that has the highest temperature in the chip is the weakest and hence that deformation should occur in this layer as observed (Fig. 7.16). However, as a result of dynamic strain ageing this is not true for values of T_{mod} in the range (approximately) 500 K $< T_{\text{mod}} <$ 700 K and for such cases the chip will generally be weaker some distance from the interface where the temperature is lower. This has led a number of workers including Shaw *et al.* (1961) and Hastings

Fig. 7.17 — Photomicrograph of chip section showing built-up edge (chip was broken during mounting): 0.2% carbon steel; $\alpha = -5°$; $U = 25$ m/min; $t_1 = 0.125$ mm.

et al. (1974) to suggest that when dynamic strain ageing occurs at the interface a built-up edge will be formed by the stronger material adjacent to the tool cutting edge. For the present materials this means that there should be no built-up edge when $T_{mod} > 700$ K. By using this criterion with the results in Fig. 7.13, in which filled-in symbols have been used to indicate experimental conditions which gave a built-up edge, it was found that the cutting speeds corresponding to $T_{mod} = 700$ K (indicated by arrows with a bar on top) were close to the speeds at which built-up edge was observed to disappear for $t_1 = 0.5$ mm and also in some cases for $t_1 = 0.25$ mm. However, for $t_1 = 0.125$ mm the speed required to reach the required T_{mod} was much greater than the speed at which built-up edge disappeared. For such cases it was found that the average temperature T_{int} at the interface corresponding to this latter speed was in the range from 710 to 730°C. Because of this it was decided by Hastings *et al.* (1980) to introduce a second criterion, namely that even if $T_{mod} < 700$ K there would be no

Sec. 7.4] Comparison of predicted and experimental chip geometry

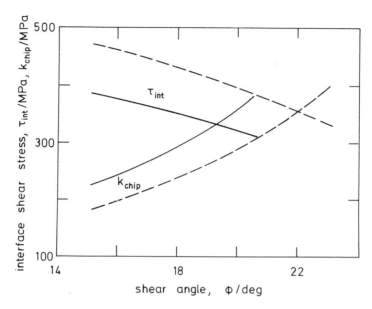

Fig. 7.18 — Curves of shear stress at tool–chip interface giving solution point values of shear angle: ———, 0.2% carbon steel; – – –, 0.38% carbon steel; $\alpha = -5°$; $U = 400$ m/min; $t_1 = 0.25$ mm.

built-up edge if $T_{int} > 1000$ K and the speeds corresponding to this are indicated in Fig. 7.13 by the arrows without a bar. It can be seen that by using these two criteria the end of the built-up edge range can be predicted exceptionally well. It is interesting to note that for the work materials used the temperature at which austenitisation starts at normal heating rates is approximately 700°C. From this it might be speculated that if the time available allowed sufficient austenite to be formed to reduce the volume of ferritic material available for dynamic strain ageing significantly then this could explain the disappearance of built-up edge at temperatures of this order. However, this would seem to be out of the question when the extremely short time available in machining is considered.

It is interesting to note that the values of C obtained from the calculations varied from 3.05 to 4.98 with an average value of 4.13 for the 0.2% carbon steel and from 4.06 to 7.34 with an average of 5.82 for the 0.38% carbon steel. No measurements were made by Hastings *et al.* (1980) in their experiments of the strain-rates in the chip formation zone for comparison with these values but it can be seen that they are of the same order of magnitude as the value of 5.9 found from the strain-rate results of Stevenson and Oxley given in Fig. 7.2. There is also some support for the predicted increase in C and hence strain-rate with increase in carbon content from the experimental results of Stevenson *et al.* (1978) who measured strain-rates in the chip formation zone using a similar method to that of Stevenson and Oxley (1969–1970).

Further comparisons between predicted and experimental results have been made by Kopalinsky and Oxley (1984) who carried out a comprehensive study of the size effect in machining. They made turning tests under approximately plane strain conditions for $\alpha = -5°$, $-25°$ and $-50°$ for t_1 values in the range 0.006–0.2 mm on a plain carbon steel of chemical composition 0.48% C, 0.3% Si, 0.13% S, 0.8% Mn, 0.019% P. The cutting speed used was 420 m/min, thus ensuring the absence of built-up edge in the tests. Special care was taken in the tests to ensure that the radius on the tool cutting edge was much smaller than the smallest value of t_1 used. Experimental values of specific cutting pressure ($F_C/t_1 w$), ϕ and k_{AB} are given in Figs 7.19 and 7.20

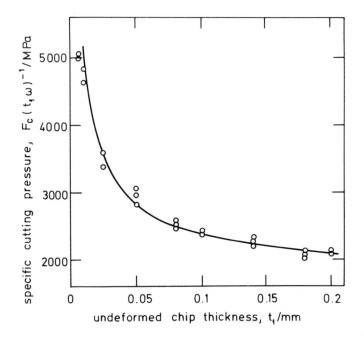

Fig. 7.19 — Predicted and experimental specific cutting pressure results: $\alpha = -5°$; $U = 420$ m/min.

for $\alpha = -5°$. The values of k_{AB} were found by dividing the component of measured forces along the plane AB, F_S (Fig. 7.1), by the area of AB defined by the experimental values of ϕ. The results in Fig. 7.19 show that the specific cutting pressure more than doubles if t_1 is reduced from 0.2 mm to 0.01 mm. A number of attempts have been made to explain why the specific cutting pressure should increase with decrease in t_1 in this way. Backer et al. (1952) reasoned that as t_1 is reduced the probability of finding dislocation sources is also reduced and as a result the flow stress and hence specific cutting pressure will both increase. Larson-Basse and Oxley (1973) considered it to be a strain-rate effect, a reduction in t_1 increasing $\dot{\gamma}_{AB}$ (see equation (6.1)) and hence increasing k_{AB} and specific cutting pressure. Looking for less sophisticated reasons for the size effect Armarego and Brown (1962) suggested that the profile of the tool cutting edge might be an important factor. Thus the radius on the cutting edge of even a 'sharp' tool and the scar which occurs on the clearance

Sec. 7.4]   **Comparison of predicted and experimental chip geometry**   133

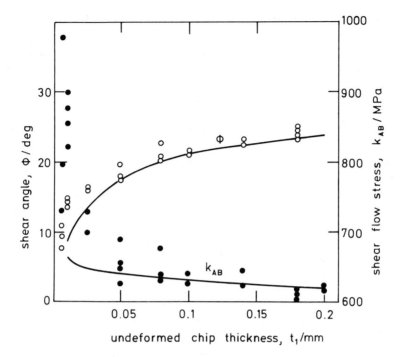

Fig. 7.20 — Predicted and experimental shear angle and shear flow stress results: $\alpha = -5°$; $U = 420$ m/min.

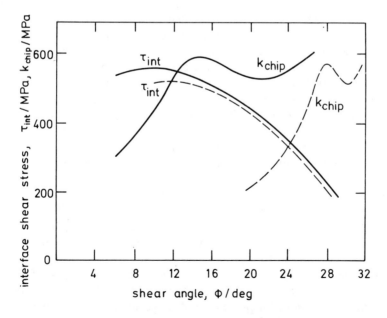

Fig. 7.21 — Curves of shear stress at tool–chip interface giving solution point values of shear angle: $\alpha = -5°$; $U = 420$ m/min; ———, $t_1 = 0.02$ mm; – – –, $t_1 = 0.2$ mm.

face of the tool as a result of wear might both be expected to contribute to the observed increase in specific cutting pressure with decrease in t_1 by making a proportionally greater contribution to the measured cutting forces at smaller values of t_1. Kopalinsky and Oxley (1984) contended that in their experiments the increase in specific cutting pressure with decrease in t_1 could mainly be accounted for by the accompanying decrease in ϕ (Fig. 7.20). They explained this decrease in ϕ in terms of the machining theory in the following way.

The lines given in Figs. 7.19 and 7.20 represent predicted results obtained from the machining theory. (Kopalinsky and Oxley used the machining theory in the form represented by the chart in Fig. 7.12. In their calculations they assumed material properties appropriate to the 0.48% carbon steel and took both temperature factors η and ψ equal to unity as this gave the best fit between predicted and experimental results.) It can be seen that good agreement exists between the predicted and experimental values of specific cutting pressure in Fig. 7.19. An attempt can therefore be made with some justification to explain the size effect by reference to the results predicted from the machining theory. Thus, the increase in specific cutting pressure with decrease in t_1 can be attributed mainly to the accompanying decrease in ϕ (Fig. 7.20) and consequently less than proportional decrease in F_C with decrease in t_1. In addition, the predicted increase in k_{AB} with decrease in t_1 (Fig. 7.20) will also make a slight contribution. The decrease in ϕ with decrease in t_1 can be explained by considering the results in Figs. 7.21 and 7.22. Fig. 7.21 shows that while there is little difference in the curves of τ_{int} for $t_1 = 0.02$ mm and $t_1 = 0.2$ mm the value of k_{chip} at the intersection point for $t_1 = 0.02$ mm is much higher than for $t_1 = 0.2$ mm and consequently the solution point value of ϕ is much lower. The reason for this increase in k_{chip} can be found in Fig. 7.22 which shows that the decrease in T_{int} with decrease in t_1 is sufficiently large to give a reduction in T_{mod} in spite of the accompanying decrease in $\dot{\gamma}_{int}$. In fact T_{mod} reduces from 835 K at $t_1 = 0.2$ mm to 682 K at $t_1 = 0.02$ mm which causes k_{chip} to increase from 328 MPa to 537 MPa at the solution points (Fig. 7.21) and as a result ϕ falls from 24° to 12.2°. With regard to k_{AB}, the results in Fig. 7.23 show that $\dot{\gamma}_{AB}$ and T_{AB} both increase with decrease in t_1; however they interact in such a way that T_{mod} changes little with t_1. The predicted slight increase in k_{AB} with decrease in t_1 (Fig. 7.20) therefore results mainly from the corresponding increase in shear strain γ_{AB} (Fig. 7.23). It can be concluded on the basis of the predicted results that the decrease in T_{int} with decrease in t_1 and the corresponding changes in k_{chip} are the most important factors in explaining the size effect. However, while the experimental values of ϕ and k_{AB} (Fig. 7.20) support the above reasoning for $t_1 > 0.05$ mm there are marked discrepancies between experimental and predicted results for smaller t_1 values. Indeed, the experimental results for $t_1 < 0.05$ mm strongly suggest that the increase in specific cutting pressure in this range results mainly from an increase in k_{AB} and that the decrease in ϕ is less than predicted. For $t_1 < 0.05$ mm, $\dot{\gamma}_{AB}$ (Fig. 7.23) is predicted to increase rapidly and it is possible that in this range the method used to represent the flow stress properties as functions of T_{mod} is inadequate as it does not take sufficient account of the possible increased sensitivity of flow stress to strain-rate at very high strain-rates. This could become extremely important in abrasion processes such as grinding where t_1 can be very small and requires further investigation.

Sec. 7.4] Comparison of predicted and experimental chip geometry 135

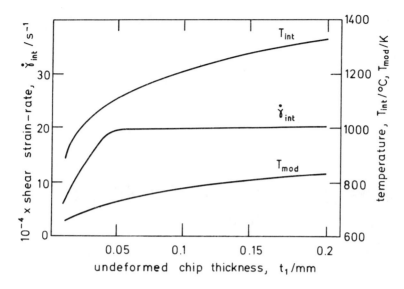

Fig. 7.22 — Predicted temperatures and strain-rates at tool–chip interface: $\alpha = -5°$; $U = 420$ m/min.

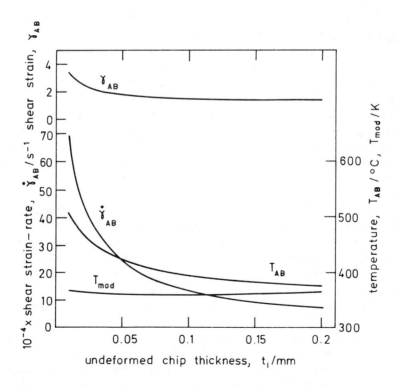

Fig. 7.23 — Predicted temperatures and strain-rates at AB: $\alpha = -5°$; $U = 420$ m/min.

8

Predicting cutting forces for oblique machining conditions

8.1 INTRODUCTION

Although many practical machining processes are approximately orthogonal they could often be represented more accurately by a chip formation model in which the cutting edge is not set normal to the cutting velocity. In Fig. 8.1 the angle i, measured between the cutting edge and the normal to the cutting velocity in the plane of the newly machined surface, is known as the inclination angle. When $i \neq 0°$ then the process is known as oblique machining. The main difference between oblique machining and orthogonal machining is that the chip flow direction is in general no longer normal to the cutting edge and the process is three dimensional. Because of the extreme difficulty of analysing such three-dimensional plastic flow problems it is not surprising that to date there have been no oblique machining analyses comparable with those for the orthogonal case. However, as will be seen, a number of attempts have been made to develop approximate methods for predicting the forces in oblique machining.

It is well known from experiment that the slope of the tool cutting face relative to the cutting velocity is one of the most important parameters in determining cutting forces, etc. In orthogonal machining this slope is defined by a single angle, the rake angle α (Fig. 1.2), where α is the angle between the cutting face and the normal to the cutting velocity measured in the plane normal to the cutting edge. In oblique machining it is not obvious which slope should be considered. For example, it could be the angle between the cutting face and the normal to the cutting velocity measured in the plane containing the cutting velocity and chip velocity which is sometimes called the effective rake angle α_e. Alternatively, it could be the so-called normal rake angle α_n measured in the plane normal to the cutting edge as in orthogonal machining. Experiments made by Brown and Armarego (1964) in which α_n and α_e were in turn held constant while the inclination angle i was varied showed little variation in the cutting force in the cutting velocity direction for constant α_n but large

Sec. 8.2] **Determining the chip flow direction** 137

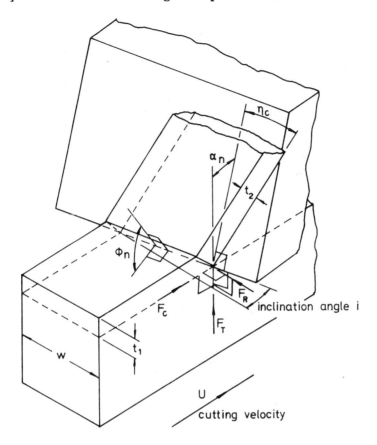

Fig. 8.1 — Oblique chip formation model.

variation for constant α_e. From these results, therefore, it might be concluded that α_n is the rake angle most directly related to the process. In addition to deciding which is the most appropriate rake angle it is clear that an essential requirement for predicting cutting forces for oblique machining conditions is a knowledge of the chip flow direction. Various equations which have been proposed for determining this direction are now considered. It should be noted that in this chapter attention is still limited to cutting on a single, straight cutting edge.

8.2 DETERMINING THE CHIP FLOW DIRECTION

Kronenberg (1954) based his definition of true rake angle in oblique machining on an intuitive feeling that the chip flow direction would be in the plane parallel to the cutting velocity and perpendicular to the machined surface. This yielded the relationship

$$\tan \eta_c = \tan i \sin \alpha_n \tag{8.1}$$

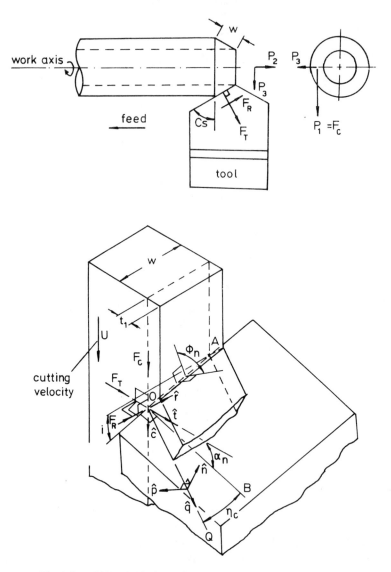

Fig. 8.2 — Oblique chip formation model referred to tube turning process.

where, referring to Fig. 8.1, η_c is the chip flow angle measured between the chip velocity and the normal to the cutting edge in the cutting face of the tool, i is the inclination angle and α_n is the normal rake angle. There is little if any experimental support for equation (8.1). Merchant (1944), assuming that the shear strain in the chip would be a minimum, derived an expression showing that to satisfy this condition the chip flow angle would need to be equal to the inclination angle. Stabler (1951) found the same relationship from experimental results, namely

$$\eta_c = i \tag{8.2}$$

Sec. 8.2] Determining the chip flow direction

which is now generally known as Stabler's flow rule. This flow rule has been shown to give good agreement with experimental results by many workers including Pal and Koenigsberger (1968) and has been widely accepted as a satisfactory predictor of chip flow direction. Later Stabler (1964) presented further experimental data which led him to modify the above relation to

$$\eta_c = ki \qquad (8.3)$$

where k is between 0.9 and 1.0 and varies with the work material and cutting conditions. Experiments made by Russel (1964) showed that η_c varied significantly with normal rake angle α_n. To allow for this Russel and Brown (1966) proposed the equation

$$\tan \eta_c = \tan i \cos \alpha_n \qquad (8.4)$$

which they showed gave a good fit with Russel's experimental results. In Russel's experiments the chip flow angle was found to be invariable with the cutting speed. However, this was not found to be the case in experimental observations made by Zorev (1966), who suggested the relation

$$\eta_c = \frac{i}{U^{0.08}} \qquad (8.5)$$

where U is the cutting speed in m/min, to account for the effect of cutting speed.

The equations so far considered are relatively simple and require a knowledge of only given cutting conditions such as i, α_n and U in order to predict η_c. More complex equations which in addition to given cutting conditions require a knowledge of factors which must either be predicted from machining theory or measured experimentally have also been proposed. For example, Armarego and Brown (1969) using a shear plane model of chip formation (Fig. 8.1) and assuming (i) that the resultant shear force on the shear plane acted in the resultant shear velocity direction and (ii) that the resultant frictional force on the tool face acted in the direction of chip flow, showed from geometry that

$$\tan \eta_c = \frac{\tan i \cos \alpha_n}{\tan(\phi_n + \lambda_n)} + \tan i \sin \alpha_n \qquad (8.6)$$

where ϕ_n (Fig. 8.1) is the shear angle measured in the normal plane and λ_n is the mean friction angle at the tool–chip interface measured in the same plane. An equivalent relationship was derived by Luk (1972) from the same assumptions. Lin and Oxley (1972) using experimental values of λ_n together with values of ϕ_n predicted from orthogonal machining theory showed that equation (8.6) predicted values of η_c in good agreement with their experimental values measured from wear marks on the tool cutting face. However, it should be noted as pointed out by Lin and

Oxley that their experimental results for η_c gave an even better fit with equation (8.2).

8.3 CUTTING FORCE PREDICTION METHODS

In one of the earliest attempts to predict cutting forces for oblique machining conditions Lin and Oxley (1972) assumed that as an approximation the flow in the normal plane could be treated as plane strain deformation with orthogonal machining theory used to predict the corresponding shear angle and cutting force values. To relate the flow in the normal plane to the complete (oblique) flow it was assumed that the shear plane model of oblique chip formation given in Fig. 8.1 could be used to represent the overall chip and force geometry in the same way that the shear plane model given in Fig. 2.1 is assumed to represent this geometry in developing the orthogonal theory described in the previous chapter. In this way, making the same assumptions as made by Armarego and Brown (1969) in deriving equation (8.6), Lin and Oxley obtained equations for the three components of cutting force F_C, F_T and F_R (Fig. 8.1) in terms of normal plane values of shear angle, etc., with η_c given by equation (8.6). In making predictions an early form of the orthogonal theory described in the previous chapter was used which required the average friction angle at the tool–chip interface to be included in the given information. Also, the work material flow stress properties used were obtained from machining and not independent tests. The agreement between predicted and experimental values of chip flow direction and cutting forces was found to be good. Later Lin (1978) used essentially the same approach but following Fenton and Oxley (1968–1969) replaced the average friction angle by the shear strength of the chip at the tool–chip interface as the friction parameter, thus making the theory more truly predictive, although he still used work material flow stress properties obtained from machining tests.

Usui and his co-workers (Usui *et al.* 1978a,b, Usui and Hirota 1978) have made a notable attempt to provide a comprehensive oblique machining theory for predicting not only chip geometry and cutting forces but also cutting temperatures and stresses, and in this way tool wear rates. In this work the flow in the plane containing the cutting velocity and chip velocity was treated as equivalent to plane strain deformation with the corresponding shear angle, mean friction angle and work material shear strength obtained from experimental orthogonal machining results. The chip flow direction and hence the three components of cutting force were determined by assuming that the appropriate value of η_c would be that which minimised the energy of chip formation. Predictions were made for various shapes of cutting tool including those with a nose radius, and considering the approximate nature of the analysis the agreement between predicted and experimental results was excellent.

Lin *et al.* (1982) noted that all of the existing methods for predicting cutting forces for oblique conditions required that machining tests had first to be made in order to obtain data needed in making predictions. To overcome this clearly unsatisfactory situation they developed a method for predicting these forces using the orthogonal theory described in the previous chapter as a basis. One method of making predictions would have been to use the orthogonal theory to determine values of shear angle, friction angle and work material shear strength, and then to assume that these values applied to the plane containing the cutting velocity and chip velocity and to use these in the way suggested by Usui et al. (1978a,b) and Usui and Hirota (1978) to determine the chip flow direction and the three components of cutting force. Alternatively, a similar approach based on the analysis of Lin and Oxley (1972) could have been used. However, considering the complexity of oblique chip formation and the simplifying assumptions involved, Lin *et al.* (1982) decided that these analyses were somewhat oversophisticated and that a simpler, semi-empirical method of the kind described by Armarego and Brown (1969) would be more appropriate. This makes use of the experimental observations (i) that for a given normal rake angle α_n and other cutting conditions, the forces F_C and F_T (Fig. 8.1) are very nearly independent of inclination angle i, and (ii) that the chip flow direction approximately satisfies Stabler's flow rule, i.e. equation (8.2). The method developed by Lin *et al.* (1982) for predicting forces which forms the basis for all the predictions made in this book for oblique conditions is now described.

In the light of observation (i) above it is assumed that F_C and F_T can be determined from the orthogonal theory by taking $i=0°$ irrespective of its actual value with $\alpha=\alpha_n$. Then, using these values of F_C and F_T, the third component F_R (Fig. 8.1) can be found for the given value of i, with η_c given by equation (8.2), from the necessary condition that the resultant cutting force R must lie in the plane normal to the tool cutting face and containing the resultant frictional force acting in the chip flow direction. By applying vector analysis Young (1986) has shown that an expression for F_R which satisfies this condition can be obtained in the following way. In Fig. 8.2 a cartesian coordinate system is chosen such that the unit vectors $\hat{c}$, $\hat{t}$ and $\hat{r}$ are along the three mutually perpendicular force component directions F_C, F_T and F_R respectively. Therefore the resultant cutting force R in vector form is

$$R = F_C\hat{c} + F_T\hat{t} + F_R\hat{r} \tag{8.7}$$

Based on this coordinate system, the unit vector $\hat{a}$ in the direction of the cutting edge OA is expressed as

$$\hat{a} = -\sin i\,\hat{c} + \cos i\,\hat{r} \tag{8.8}$$

and the unit vector $\hat{b}$ along the line OB, the intersection between the normal plane and tool cutting face plane, as

$$\hat{b} = \sin\alpha_n \cos i\,\hat{c} + \cos\alpha_n\,\hat{t} + \sin\alpha_n \sin i\,\hat{r} \tag{8.9}$$

The unit vector $\hat{q}$ along the chip flow direction OQ is given by

$$\hat{q} = -\sin\eta_c\,\hat{a} + \cos\eta_c\,\hat{b} \tag{8.10}$$

which on substituting for $\hat{a}$ and $\hat{b}$ from equations (8.8) and (8.9) gives

$$\hat{q} = (\sin\alpha_n \cos i \cos\eta_c + \sin i \sin\eta_c)\hat{c} + \cos\alpha_n \cos\eta_c\,\hat{t} + (\sin\alpha_n \sin i \cos\eta_c - \cos i \sin\eta_c)\hat{r} \tag{8.11}$$

The unit vector $\hat{n}$ normal to the tool cutting face plane is found from the cross product of two independent unit vectors on this plane, for example $\hat{a}$ and $\hat{b}$, which gives

$$\hat{n} = \hat{b} \times \hat{a}$$
$$= \cos\alpha_n \cos i\, \hat{c} - \sin\alpha_n\, \hat{t} + \cos\alpha_n \sin i\, \hat{r} \tag{8.12}$$

Similarly,

$$\hat{p} = \hat{q} \times \hat{n}$$
$$(\sin i \cos\eta_c - \sin\alpha_n \cos i \sin\eta_c)\hat{c} - \cos\alpha_n \sin\eta_c\, \hat{t} -$$
$$(\cos i \cos\eta_c + \sin\alpha_n \sin i \sin\eta_c)\hat{r} \tag{8.13}$$

where $\hat{p}$ is the unit vector normal to the plane in which the resultant cutting force R lies. Since the vector $\hat{p}$ and the resultant force R are orthogonal, their inner product (or scalar product) is zero, that is

$$\hat{p}\cdot R = 0 \tag{8.14}$$

Substituting for R and $\hat{p}$ from equations (8.7) and (8.13) respectively and rearranging the similar terms in the equation gives the relation

$$F_R = \frac{F_C(\sin i - \cos i \sin\alpha_n \tan\eta_c) - F_T \cos\alpha_n \tan\eta_c}{\sin i \sin\alpha_n \tan\eta_c + \cos i} \tag{8.15}$$

To check the accuracy of their predicted results Lin *et al.* (1982) used a turning test with the tool being fed in a direction parallel to the axis of rotation of the work (Fig. 8.2). A tube workpiece was used so that cutting was only on one edge. To relate the three components of cutting force predicted as above to the cutting forces normally measured in turning the chip formation model used in the analysis (Fig. 8.1) has to be related to the actual process used. In the tests two cases were considered. In the first, the plane containing the cutting edge and cutting velocity was normal to the feed direction. That is, the side cutting edge angle C_s (Fig. 8.2), which is measured in the plane normal to the cutting velocity, was zero. It can be seen from Fig. 8.2, in which the chip formation model in Fig. 8.1 is turned through 90° to make it align with the actual turning arrangement, that for $C_s = 0°$ the forces F_C and F_T act in the directions parallel to the cutting and feed velocities, with F_R acting in the radial direction. In applying the orthogonal theory ($i = 0°$) to this case the undeformed chip thickness, t_1, is equal to the axial feed measured in mm/rev, and the width of cut, w, is equal to the tube wall thickness. In the second case, the plane containing the cutting edge and cutting velocity was rotated so that $C_s \neq 0°$ and in applying the orthogonal theory to this case $t_1 = \text{feed} \times \cos C_s$ and $w = (\text{tube wall thickness})/\cos C_s$. The forces F_T and F_R now no longer act in the feed and radial directions and it is usual to redefine the forces acting on the tool as (say) P_1, P_2 and P_3 where these act in the cutting, feed and radial directions respectively. If as in Fig. 8.2, and as would usually be the case, the sense of both the side cutting edge angle C_s and inclination angle i is to make the chip flow away from the axis of rotation of the work then P_1, P_2 and P_3 can be expressed in terms of F_C, F_T and F_R by the relations

Sec. 8.4] **Comparison of predicted and experimental cutting force results** 143

$$P_1 = F_C$$
$$P_2 = F_T \cos C_s + F_R \sin C_s \qquad (8.16)$$
$$P_3 = F_T \sin C_s - F_R \cos C_s$$

where the positive directions P_1, P_2 and P_3 are taken as the cutting velocity, negative feed and radially outward directions.

The next section considers a comparison made by Lin et al. (1982) of forces predicted in the above way with experimentally measured forces.

468.4 COMPARISON OF PREDICTED AND EXPERIMENTAL CUTTING FORCE RESULTS

In obtaining the predicted results for given values of normal rake angle α_n, cutting speed U, inclination angle i, side cutting edge angle C_s, feed and tube wall thickness, the following procedure is used. The values of t_1 and w, which depend on C_s as described above, are determined and then taking $\alpha = \alpha_n$ the orthogonal theory is used to find F_C and F_T, the method of calculation being as summarised in Fig. 7.12. Once F_C and F_T are known then, for the given value of i, F_R is found from equation (8.15) with η_c given by equation (8.2). The work materials used in the experiments were steels of chemical composition 0.44% C, 0.016% P, 0.87% Mn, 0.32% Si, 0.032% S, 0.95% Cr, 0.2–0.3% Cu, <0.1% Ni and 0.58% C, 0.105% Si, 0.947% Mn, 0.08% Cr. The values of σ_1 and n needed in the calculations were obtained from the equations given in Appendix A3 and the thermal properties S and K from equation (6.18) and equations (6.19) and (6.20) respectively. The temperature factors η and ψ were both taken as 0.7. Results found in this way for zero side cutting edge angle for the 0.58% carbon steel are represented by the lines in Fig. 8.3. When $C_s \neq 0°$ then the forces P_1, P_2 and P_3 are found from equations (8.16) using the values of F_C, F_T and F_R determined as above. Results for the 0.44% carbon steel showing the effect of side cutting edge angle are represented by the lines in Fig. 8.4. The experimental force results for the 0.58% and 0.44% carbon steels given in Figs. 8.3 and 8.4 were measured by Lin et al. (1982) from turning tests made on tubes of these materials. The lathe and force measuring system used in the tests were the same as those described in section 7.4. Average values of force for each cutting condition are given, the fluctuation in forces being less than 10% in the absence of built-up edge which was the case for all except the slowest cutting speeds (less than 40 m/min). Brazed-on tungsten carbide tipped cutting tools were used. The rake angles, cutting speeds, etc., used in the tests are as indicated in Figs. 8.3 and 8.4.

The experimental results in Fig. 8.3(a) show that although there is a great deal of scatter there is no obvious trend in F_C and F_T with inclination angle i, thus supporting the earlier experimental findings of Armarego and Brown (1969). The predicted and experimental results show reasonably good agreement for all three components of force except for the lower cutting speeds, with F_C and F_T decreasing with increase in α_n and U, and F_R (Fig. 8.3(b)) decreasing with increase in α_n and increasing with increase in U and i. The predicted and experimental results in Fig. 8.4 which show the influence of side cutting edge angle C_s are also in reasonable agreement although there are obvious discrepancies which a less approximate method might be expected to account for. P_1, P_2 and P_3 can be seen to vary with α_n and U in much the same way

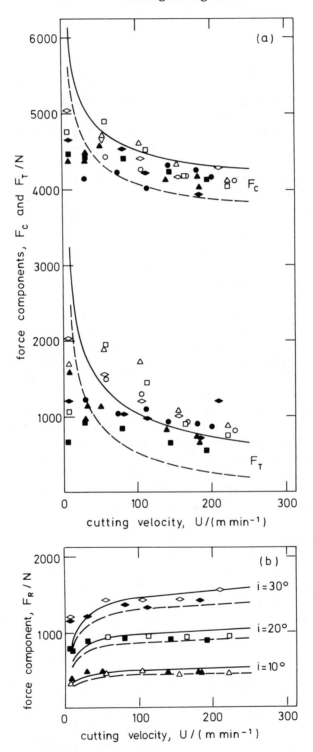

Fig. 8.3 — Predicted and experimental cutting forces for zero side cutting edge angle: ———, α_n = 9°; – – –, α_n = 15°C; ○, ●, i = 0°; △, ▲, i = 10°; □, ■, i = 20°; ◇, ◆, i = 30°; open symbols, α_n = 9°; filled-in symbols, α_n = 15° (t_1 = 0.5 mm; w = 5.0 mm; 0.58% carbon steel).

Sec. 8.4] Comparison of predicted and experimental cutting force results

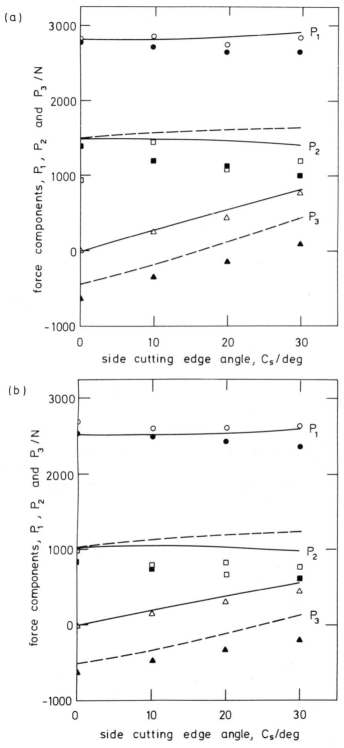

Fig. 8.4 — Predicted and experimental cutting forces showing effect of side cutting edge angle: ———, $i=0°$; – – –, $i=20°$; open symbols $i=0°$; filled-in symbols $i=20°$ ($f=0.25$ mm/rev; wall thickness = 5.0 mm; 0.44% carbon steel): (a) $\alpha_n=0°$; $U=100$ m/min; (b) $\alpha_n=0°$; $U=300$ m/min.

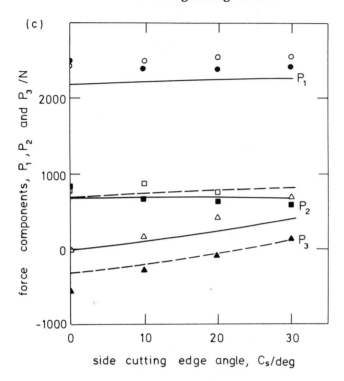

Fig. 8.4 (cont.) — Predicted and experimental cutting forces showing effect of side cutting edge angle: ———, $i = 0°$; – – –, $i = 20°$; open symbols $i = 0°$; filled-in symbols $i = 20°$ ($f = 0.25$ mm/rev; wall thickness = 5.0 mm; 0.44% carbon steel): (c) $\alpha_n = 15°$; $U = 100$ m/min.

as for the results in Fig. 8.3 and it is the changes in P_3 with C_s and i which are of most interest. It is predicted that an increase in C_s will increase P_3 while an increase in i will decrease P_3 and for $i = 20°$ it can be seen that P_3 is predicted to be negative (i.e. acting radially inwards), only becoming positive for the larger values of C_s and the experimental results confirm this.

9

Allowing for end cutting edge effects in predicting cutting forces in bar turning with oblique machining conditions

9.1 INTRODUCTION

The experiments of Lin *et al.* (1982) described in section 8.4 were made on tubes to ensure that cutting took place only on one edge as assumed in the method developed for predicting forces. In practice cutting tools are more complex and are normally seen to have two edges that cut simultaneously, e.g. the conventional single-point lathe tool. Noting this, Hu *et al.* (1986) made a series of oblique turning tests on a bar to see how the cutting which would occur on the end (secondary) cutting edge in addition to cutting on the side (main) cutting edge would influence the cutting forces and chip flow direction. It was then hoped to modify the method for predicting forces to account for any observed influence of the end cutting edge. This investigation is now described.

9.2 EXPERIMENT

Bar turning tests were made with the three force components P_1, P_2 and P_3 (Fig. 8.2) in the cutting, feed and radial directions respectively, measured using a three-component dynamometer. The lathe and force measuring system used in the tests were the same as those described in section 7.4. The cutting conditions used were as follows: normal rake angle $\alpha_n = 5°$, $9°$, $9.8°$, $10°$, $10.5°$, $15°$ and $20°$; side cutting edge angle $C_s = 0°$, $10°$, $20°$ and $30°$; end cutting edge angle $C_e = 6°$; inclination angle i in the range from $-9°$ to $10.5°$; feed $f = 0.125$, 0.25, 0.355 and 0.5 mm/rev; depth of cut $d = 4.0$ mm; cutting speed $U = 150$ m/min. The tests were limited to one (typical) cutting speed because the main interest in the investigation was the influence of tool angles and feed on the cutting forces and chip flow direction. The speed selected was such as to ensure the absence of built-up edge in the tests. The work material used in the tests was a steel of chemical composition 0.19% C, 0.88% Mn, 0.27% Si, 0.085% Cr, 0.021% P, 0.02% S, 0.02% Ni, 0.02% Cu, 0.02% Ti, 0.005% Mo. Brazed-on tungsten carbide tipped tools were used in all the tests. The tools were carefully prepared (ground and lapped) so that the tool nose radius was negligible compared with the feed, even for the smallest feed used. This was to minimise nose radius effects which are considered in the next chapter.

A number of techniques were tried for measuring the chip flow direction (angle η_c in Fig. 8.2) and it was found that the most accurate method was to take still photographs of the chip flowing over the tool cutting face while actually machining. To achieve this a special, rigid stand was designed and built which was fixed on the rear tool post which moved with the front tool post (which held the dynamometer) during machining. A 35 mm camera was connected with a universal attachment to the stand and during a test the lens axis of the camera was adjusted so that it was normal to the cutting face of the tool. With this arrangement the camera moved with the tool during a test and the resulting photographs were taken in the correct plane. A strong light source was used to increase the illumination and thus to improve the contrast in the observed region. A typical photograph obtained in this way showing a 'frozen' view of the chip flowing over the tool is given in Fig. 9.1. Care was taken to

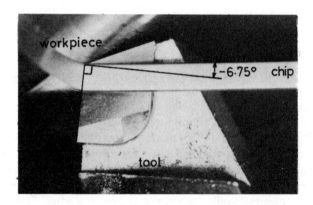

Fig. 9.1 — Photograph of chip flowing over tool cutting face.

measure the chip flow angle η_c at the actual cutting edge; this assumed greater importance when the chip was less straight than shown in Fig. 9.1. Experimental cutting force (P_1, P_2 and P_3) and chip flow angle (η_c) values are given in Figs. 9.2 and 9.3. To ensure that the wear of the cutting edge did not have a large effect on the measured force and chip flow angle values frequent checks were made of the wear, and the tool was reground and lapped if the wear became significant.

9.3 PREDICTED RESULTS NEGLECTING END CUTTING EDGE EFFECTS

In calculating forces the method of Lin *et al.* (1982) described in section 8.3 for cutting on a single cutting edge was used with the depth of cut equivalent to the tube

Sec. 9.3] Predicted results neglecting end cutting edge effects

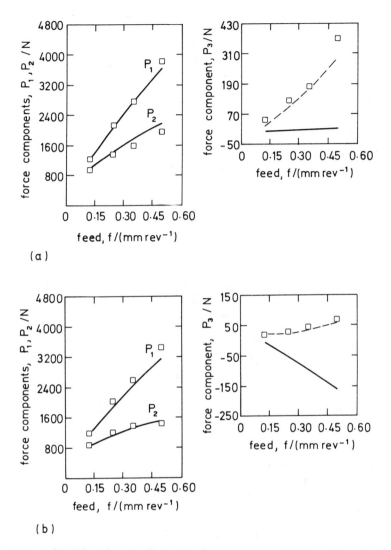

Fig. 9.2 — Predicted and experimental cutting forces: ———, predicted results neglecting end cutting edge effects; – – –, predicted results using equivalent cutting edge ($d=4.0$ mm; $U=150$ m/min; $C_e=6°$): (a) $\alpha_n=5°$; $C_s=0°$; $i=-0.5°$; (b) $\alpha_n=10.5°$; $C_s=1°$; $i=10.5°$.

wall thickness of the previous calculations. Work material properties appropriate to the chemical composition of the work material used in the tests were obtained as before from the equations given in Appendix A3 and in section 6.3. The chip formation zone temperature factor η was again taken as 0.7 but the tool–chip interface factor ψ was calculated from the empirical equation

$$\psi=(0.776-0.75t_1-0.3t_1^2+1.37t_1^3+(0.014\alpha_n+0.93) \tag{9.1}$$

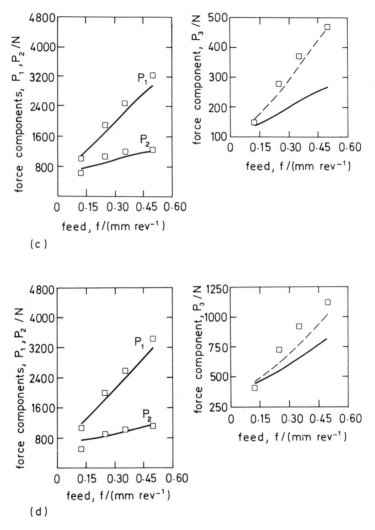

Fig. 9.2 (cont.) — Predicted and experimental cutting forces: ———, predicted results neglecting end cutting edge effects; - - -, predicted results using equivalent cutting edge ($d=4.0$ mm; $U=150$ m/min; $C_e=6°$): (c) $\alpha_n=15°$; $C_s=10°$; $i=-3.2°$; (d) $\alpha_n=15°$; $C_s=30°$; $i=-9°$.

where α_n is measured in degrees and t_1 in millimetres. This equation allows for the influence of t_1 on temperature in the manner suggested by Mathew *et al.* (1979) and also allows for the decreased heat sink capacity of the tool and hence higher temperature with an increase in rake angle as suggested by Hu and Mathew (1983). For the cutting conditions considered by Hu *et al.* (1986) ψ determined in this way fell in the range 0.4–0.85.

The predicted values of P_1 and P_2 (Fig. 9.2) obtained in this way can be seen to agree well with the experimental values for most of the cutting conditions con-

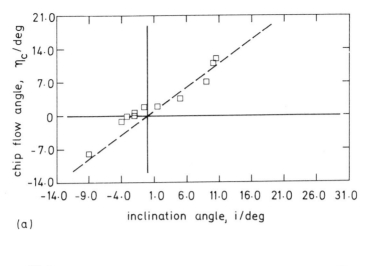
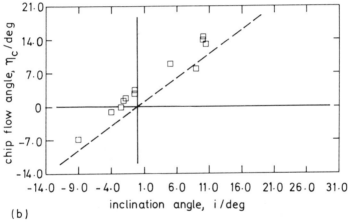

Fig. 9.3 — Comparison of chip flow angle experimental results with Stabler's flow rule $\eta_c = i$: (a) $f = 0.125$ mm/rev; (b) $f = 0.25$ mm/rev.

sidered. However, as the feed is increased the radial force P_3 is seriously underestimated. The reason for this can be clearly seen from the experimental chip flow angle results in Fig. 9.3 which show large departures from the Stabler flow rule $\eta_c = i$ (equation (8.2)) used in developing the method for predicting forces. It is the underestimation of η_c by this relation at higher feeds that leads to the underestimation of P_3 for these conditions.

9.4 DESCRIPTION OF AN EQUIVALENT CUTTING EDGE

There have been a number of attempts to allow for the influence of the cutting action of the end cutting edge in determining the direction of chip flow. Colwell (1954)

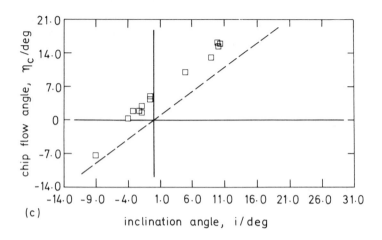

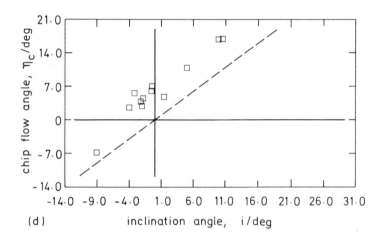

Fig. 9.3 (cont.) — Comparison of chip flow angle experimental results with Stabler's flow rule $\eta_c = i$: (c) $f=0.355$ mm/rev; (d) $f=0.50$ mm/rev.

assumed that the direction of chip flow would be perpendicular to the major axis of the projected area of cut (equivalent cutting edge), as shown in Fig. 9.4. He showed that this gave accurate predictions of the chip flow angle for zero rake angle tools (that is $i=0°$, $\alpha_n=0°$), but for more complicated tool geometries it was necessary to introduce semi-empirical corrections found by considering the forces involved in chip formation. Zorev (1966) derived an equation for the chip flow angle by considering the magnitudes of the forces acting on the side and end cutting edges which he assumed were directly proportional to the engaged lengths of these edges. This first equation was restricted to the case where i and α_n were both zero and a further geometrical term was introduced to account for the influence of these angles. Luk (1969) pointed out that Zorev had made errors of sign in his analysis and also that even his more general equation was only valid for very small effective rake angles (measured in the plane containing the cutting and chip velocities) and for $C_s=0°$. Luk corrected these errors and extended Zorev's analysis to more general tool shapes. Stabler (1964) in a slightly different approach to the problem, and following on his earlier work (Stabler 1951), suggested that the resultant chip flow direction could be found by adding vectorially the velocities at the two cutting edges. In doing this he assumed that the velocity magnitudes would be directly proportional to the engaged length of each edge, with the directions given by the appropriate form of his flow rule.

Noting that the poor agreement between the predicted and experimental radial cutting force (P_3) values at higher feeds (Fig. 9.2) resulted mainly from the departure of experimental values of η_c from equation (8.2) at higher feeds (Fig. 9.3), an attempt was initially made by Hu et al. (1986) to see whether any of the above approaches could be used to predict η_c more accurately. Unfortunately none of the suggested relations gave a satisfactory fit with the experimental values of η_c in Fig. 9.3. In an attempt to overcome this problem they decided to try an approach similar to that of Colwell (1954) and to assume that the side and end cutting edges could be replaced by an equivalent cutting edge as shown in Fig. 9.4. However, considering

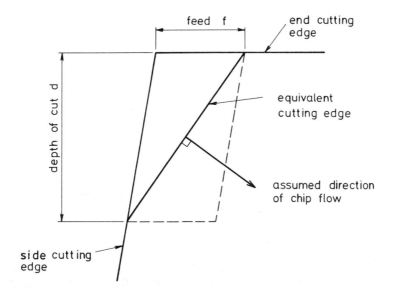

Fig. 9.4 — Colwell's (1954) model of chip flow direction.

the poor agreement between the results of chip flow angle obtained from Colwell's model and experimental results, other than for the simplified case of $i=0°$ and $\alpha_n=0°$, it was clear that in defining the equivalent cutting edge account would have to be taken of the more complicated geometry which results when $i \neq 0°$ and $\alpha_n \neq 0°$. The method used by Hu et al. in defining the equivalent cutting edge and determining the angles associated with it is now described. It is of interest that Armarego and Wiriyacosol (1978a,b) have used a somewhat similar approach in analysing the process in which a triangular-form tool cuts a V groove.

154   Allowing for end cutting edge effects   [Ch. 9

To define the geometry of a single-point tool a system of fundamental planes is adopted. As shown in Fig. 9.5 this system consists of five basic planes: P_r, P_p, P_f, P_s

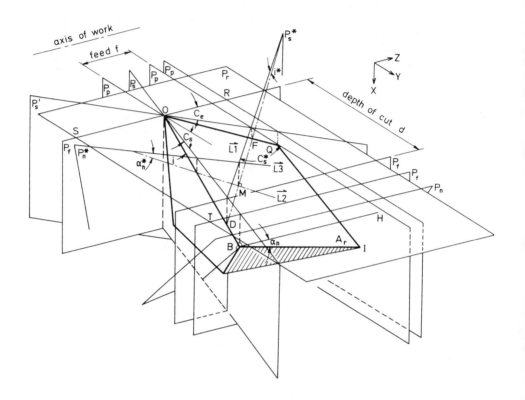

Fig. 9.5 — Model used in determining equivalent cutting edge angles.

and P_n; where P_r, the tool reference plane, is parallel to the tool base, P_p, the tool back plane, is perpendicular to the axis of the work and is along the direction of the tool axis, P_f, the assumed working plane, is parallel to the assumed direction of feed motion, P_s, the tool cutting edge plane, is tangential to the side (main) cutting edge OB and perpendicular to the tool reference plane P_r, and P_n, the cutting edge normal plane, is perpendicular to the side cutting edge. Among these planes P_r, P_p and P_f are mutually perpendicular to each other. The symbols used show that each plane is provided with a capital letter P with a subscript to denote the plane's identity. All the planes and angles defined assume that the selected point of reference lies on the side cutting edge OB. If, however, the selected point is referred to the equivalent cutting edge which is to be established and discussed below, the corresponding symbol will bear an asterisk*. For example, P_n^* symbolises the plane normal to the equivalent cutting edge.

Sec. 9.4] **Description of an equivalent cutting edge** 155

In constructing a mathematical model for the geometry of a single-point tool, which is characterised by certain basic tool angles, precise definitions of these angles are essential. These are as follows:

C_s is the side cutting edge angle; it is the acute angle that P_s makes with P_p and is measured in the reference plane, positive in the clockwise sense.

C_e is the end cutting edge angle; it is the acute angle that P_s', the end cutting edge plane, makes with P_f.

i is the inclination angle; it is the angle, measured in P_s, between the side cutting edge OB and P_r; it is positive if OB is sloping downwards from the tool tip (point O).

α_n is the normal rake angle; it is the angle measured in P_n between A_r, the cutting face plane, and P_r; it is positive if A_r is sloping downwards with respect to the side cutting edge.

η_c is the chip flow angle; it is the angle measured from the normal to the side cutting edge to the direction of chip flow on the cutting face plane; it is positive in the clockwise sense.

It is noted that all the angles shown in Fig. 9.5 are in their positive sense.

For the proposed equivalent cutting edge model, three-dimensional geometrical analysis techniques are used to determine equations for the fundamental planes and the cutting face plane of a given tool from its basic tool angles. The geometry of the tool is well defined by these planes as any of its cutting edges can be found from the intersection of two associated planes, while locating a point on a cutting edge simply means solving simultaneous equations of a plane and a line representing the cutting edge mathematically. Since the approach also requires that both the magnitude and sense of the modified angles are to be analysed, vector analysis is used. Each step of the procedure is now outlined under an appropriate heading.

(1) Basic tool geometry

To begin, a set of Cartesian coordinates is selected with the origin O at the tip of the tool (Fig. 9.5) and oriented in such a way that its X, Y, Z axes coincide with the vectors representing P_r, P_f and P_p respectively. As in the mathematical sense a plane can be extended in any direction without changing its characteristics, planes P_f and P_p with their vertical distances $ST = l$ and $OR = w$ of arbitrary length from O are then set up. (In other words, any assigned values of l and w will not affect the expression for the cutting face plane if the normalised form is used.) Following this and given values of angles C_s and i, point B (X_b, Y_b, Z_b) on the side cutting edge is located. Next the line BH is found from the intersection of P_n and P_r which contains point B. With the given normal rake angle α_n the line BI, which is on P_n as well as the cutting face plane, is determined. The point I (X_i, Y_i, Z_i) is then obtained from the intersection of the line BI and the previously selected arbitrary plane P_p passing through the point R. The equation for the cutting face plane can thus be expressed implicitly as

$$\begin{vmatrix} X & Y & Z & 1 \\ 0 & 0 & 0 & 1 \\ X_b & Y_b & Z_b & 1 \\ X_i & Y_i & Z_i & 1 \end{vmatrix} = 0$$

which on simplifying gives

$$\begin{vmatrix} X & Y & Z \\ X_b & Y_b & Z_b \\ X_i & Y_i & Z_i \end{vmatrix} = 0$$

Finally, the expression for the end cutting edge is given by solving the simultaneous equations for the cutting face plane and for P_s', which is a simple function of the end cutting edge angle C_e. In summary, the above procedure can be stated as follows: given the tool angles C_s, i, α_n and C_e, find the equations for the side cutting edge, end cutting edge and the cutting face plane.

(2) Equivalent cutting edge
The end points D and F of the equivalent cutting edge FD are determined as follows. D is found from the intersection of the side cutting edge OB and a working plane P_f at distance d (the depth of cut) from the tool tip O. F is found from the intersection of the end cutting edge OQ and a back plane P_p at a distance f (the feed) from O. Locating F in this way is an approximation and does not give the exact location of the point of intersection of the spiral workpiece surface and the end cutting edge. However, for the conditions considered in the tests the error in the position of F measured from O is less than 6%.

(3) Modified inclination angle i^*
Following the determination of the equivalent cutting edge, the corresponding cutting edge plane P_s^* can be set up. The line of intersection L1 between this plane and P_r is then found and i^* is the angle that the equivalent cutting edge FD makes with L1. The same sign convention is used as described previously.

(4) Modified side cutting edge angle C_s^*
C_s^* is identified as the angle made by L1 with the positive Y axis.

(5) Modified normal rake angle α_n^*
Since in defining α_n^* a normal plane P_n^* perpendicular to the equivalent cutting edge FD is required, an arbitrary point M on FD is selected to lie in this plane. The intersections of P_n^* with the cutting face plane and P_r are then found to be L2 and L3 respectively; α_n^* is now taken as the angle between L2 and L3.

(6) Modified chip flow angle η_c^*

Sec. 9.4] Description of an equivalent cutting edge

To relate experimentally obtained values of η_c to the equivalent cutting edge, the geometric relation $\eta_c^* = \eta_c - \angle ODF$ is used.

As would be expected, the equations for the modified parameters involve lengthy mathematical expressions in terms of the given basic tool angles and cutting conditions (feed and depth of cut) as a result of the complexity of the geometry. In most cases an explicit expression is found to be very difficult if not impossible to obtain, especially when the sense of the angles is imposed as in the present analysis. However, these calculations can be made with relative ease using a simple computer program which carries out the necessary mathematical manipulations which can be classified into three main categories: (a) defining planes such as P_r, P_f, P_p, P_s, P_n and the cutting face plane; (b) finding the intersection of two planes, or between a plane and a vector; (c) solving for the magnitude and sense of an angle made by one vector with another. Tables 9.1 and 9.2 give some typical results obtained in this way showing the influence of feed on the modified angles.

Table 9.1

Feed/(mm/rev)	C_s^*/(deg)	α_n^*/(deg)	i^*/(deg)
0.125	31.39	14.78	−9.37
0.25	32.76	14.55	−9.72
0.355	33.87	14.36	−10.00
0.5	35.38	14.09	−10.38

$C_s = 30°$; $C_e = 6°$; $\alpha_n = 15°$; $i = -9°$; $d = 4$ mm.

Table 9.2

Feed/(mm/rev)	C_s^*/(deg)	α_n^*/(deg)	i^*/(deg)
0.125	31.41	10.12	4.75
0.25	32.78	10.23	4.50
0.355	33.92	10.32	4.30
0.5	35.44	10.43	4.02

$C_s = 30°$; $C_e = 6°$; $\alpha_n = 10°$; $i = 5°$; $d = 4$ mm.

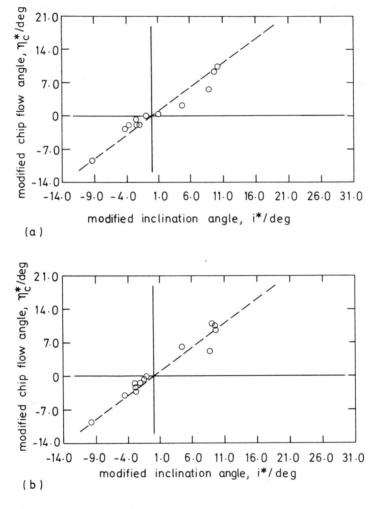

Fig. 9.6 — Comparison of chip flow angle experimental results with modified flow rule $\eta_c^* = i^*$: (a) $f=0.125$ mm/rev; (b) $f=0.25$ mm/rev.

9.5 PREDICTED RESULTS USING EQUIVALENT CUTTING EDGE

Fig. 9.6 gives the experimental chip flow angle results plotted using the modified angles η_c^* and i^*. It can be seen that Stabler's flow rule written in the form

$$\eta_c^* = i^* \tag{9.2}$$

gives excellent agreement with the experimental results. The broken lines in Fig. 9.2 represent cutting forces predicted by exactly the same method as before but with the equivalent cutting edge assumed to be the actual cutting edge. That is, equation (9.2) is used in place of equation (8.2) with $\alpha_n = \alpha_n^*$, $C_s = C_s^*$ and $i = i^*$. It can be seen that in the case of P_1 and P_2 there is no discernible difference between the two sets of

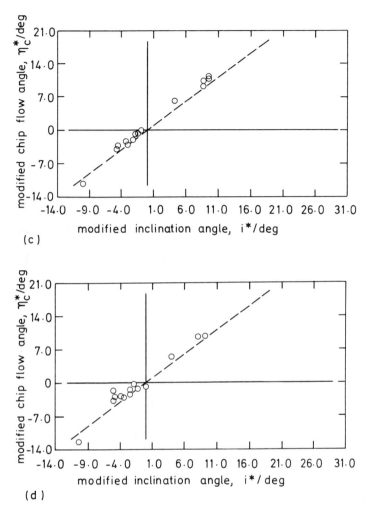

Fig. 9.6 (cont.) — Comparison of chip flow angle experimental results with modified flow rule $\eta_c^* = i^*$: (c) $f=0.355$ mm/rev; (d) $f=0.50$ mm/rev.

predicted results. However, this is not the case with the radial force P_3, and the agreement between the predicted and experimental values of P_3 can now be seen to be very good compared with the previous poor agreement.

10

Influence of tool nose radius on chip flow direction and cutting forces in bar turning

10.1 INTRODUCTION

In the previous chapter the contribution of the end cutting edge in bar turning was considered in determining the chip flow direction and the cutting forces. In the work described attention was limited to machining with sharp-nosed tools. Such tools are not typical of the tools used in industry which virtually always have a nose radius which not only improves the surface finish obtained on the machined surface but also improves the strength and wear characteristics of the tool. The development of a sufficiently accurate model for predicting the chip flow direction and cutting forces for machining operations with nose radius tools is therefore of great practical importance. The first step is clearly to make accurate predictions of chip flow direction and this is considered first.

10.2 PREDICTION OF CHIP FLOW DIRECTION

Several attempts have been made to predict the direction of chip flow when cutting with nose radius tools. Colwell (1954) suggested a simple geometric method as described in the previous chapter which assumed that the chip flow over the cutting face of the tool was perpendicular to the major axis of the projected area of cut. For the cut with nose radius tools, the major axis is in fact the segment joining the extreme points of the engaged cutting edge, as shown in Figs. 10.1(a,b,c). The equation for the predicted chip flow angle obtained from this method can be expressed in three different ways depending on the cut geometry. Fig. 10.1(a) shows the case in which the side cutting edge angle C_s is 0° and the depth of cut d is greater

(a)
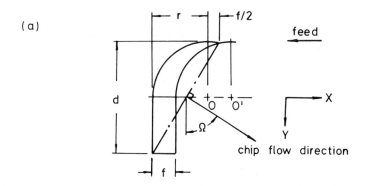

(b)
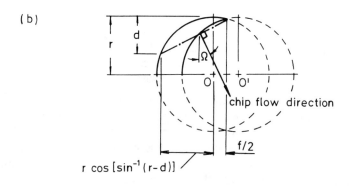

(c)
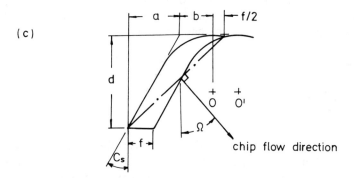

Fig. 10.1 — Colwell's (1954) method of determining the chip flow angle for nose radius tools.

than the nose radius r. From the associated geometry the relation for the chip flow angle is

$$\Omega = \cot^{-1}\left(\frac{r + f/2}{d}\right) \tag{10.1}$$

where Ω is the predicted chip flow angle measured from the tool axis (positive Y axis) and f is the feed. When the depth of cut is small enough in relation to the nose radius, so that only the curved edge is engaged (Fig. 10.1(b)), then

$$\Omega = \cot^{-1}\left[\frac{(2rd - d^2)^{1/2} + f/2}{d}\right] \quad (10.2)$$

For the more general case, where the side cutting edge angle is other than 0° and the depth of cut is greater than the nose radius, as shown in Fig. 10.1(c), a more involved relationship is obtained, namely

$$\Omega = \cot^{-1}\left(\frac{a + b + f/2}{d}\right) = \cot^{-1}\left[\frac{d \tan C_s + r \tan(\pi/4 - C_s/2) + f/2}{d}\right] \quad (10.3)$$

It should be noted that the above equations were derived based on the condition that the cutting was approximately orthogonal, i.e. both the inclination and rake angles of the nose radius tool considered were 0°. Further experimental work by Colwell led him to introduce a coefficient as a function of these two angles to account for their combined effect on the chip flow direction. This coefficient is essentially empirical and bears no apparent connection with the cut geometry.

Okushima and Minato (1959) treated the chip as a series of elements of infinitesimal width each having a tendency to flow in the direction governed by the geometry of the corresponding portion of the cutting edge. They assumed that the chip would take up a flow direction that was the average of directions of its constituents. Thus

$$\overline{\Omega} = \frac{\int \Omega(s) \, ds}{\int ds} \quad (10.4)$$

where ds is a small length of the cutting edge and $\Omega(s)$ is the local chip flow angle of the corresponding chip element leaving that portion of the edge. In this work tools with zero inclination and side cutting edge angles were considered, and each individual chip element was then assumed to flow at right angles to the corresponding part of the cutting edge. From this, the resultant chip flow angle was expressed in terms of various cutting parameters.

Usui and his co-workers (Usui et al. 1978a,b, Usui and Hirota 1978) have, as mentioned in Chapter 8, used an iterative technique to find the chip flow direction that minimised the sum of the shear energy and friction energy. In the analysis they assumed that the relationships for the effective shear angle and shear stress, each being a function of the effective rake angle, were the same as those obtained experimentally for the corresponding orthogonal cutting conditions. In this way, a typical three-dimensional cutting process, in which a nose radius tool with an inclination angle other than zero degrees was used, could be interpreted as a collection of two-dimensional cutting processes.

Other attempts to predict the chip flow angle for a cut with nose radius tools have been made and these have been reviewed by Spanns (1970) and Spanns and van Geel (1970). From experimental results they found that the resultant chip flow angle could be determined by superposition of two components; one caused by the tool nose radius and the other by the inclination angle of the side cutting edge. van Luttervelt and Pekelharing (1976) and Kluft *et al*. (1979) have analysed the effect of various cutting parameters on the chip flow direction and showed that this direction depended not only on the geometry of the tool and cutting conditions (feed, depth of cut) but also on the curvature of the workpiece, presumably because of the variation of cutting velocity across the cutting edge. The magnitude of the mean cutting velocity did not appear to affect the chip flow direction. In order to predict the chip flow angle under any combination of cutting conditions and tool geometry, they then assumed that the total effect could be found by superposition of the separate effects caused by independent parameters.

Young *et al*. (1987) made bar turning experiments with nose radius tools and measured the chip flow direction. They found that none of the above methods could predict the chip flow direction with sufficient accuracy unless use was made of empirical data, thus greatly reducing their predictive value. In view of this they introduced a new analysis of the problem by which they hoped to develop a method for predicting the chip flow direction with nose radius tools without recourse to empirical data. To simplify the problem they limited attention to tools having zero inclination angle and zero normal rake angle. Their analysis is now described in detail.

It has been reasoned by Colwell (1954) that the cutting forces are important in considering the problem of predicting the direction of chip flow. There are two force components acting on the chip at the tool–chip interface. One is the friction force F in the cutting face plane and the other is the normal force N in the direction normal to the cutting face plane. Since the normal force applies at right angles to the chip movement, it is not unreasonable to assert that the friction force is the force associated with the chip flow direction. This is the fundamental basis of the chip flow model developed by Young *et al*. In the approach used the chip is treated as a series of independent elements of infinitesimal width. The thickness and orientation of the undeformed chip section corresponding to each chip element vary. Therefore, the friction force component for each element changes in magnitude as well as direction. In what follows these friction force components are summed up to find their resultant and it is assumed that the direction of the resultant coincides with the chip flow direction. In the analysis it is assumed (i) that the elemental friction force is collinear with the local chip velocity, (ii) that the magnitude of the elemental friction force increases or decreases linearly with the local undeformed chip thickness, and (iii) that the local chip flow direction satisfies Stabler's flow rule which for $i = 0°$ means that the flow is normal to the cutting edge for the element considered. In support of the second assumption Young *et al*. have cited experimental results of Brown and Armarego (1964) which show that for cutting tests with straight-edged tools (uniform undeformed chip thickness) the friction force varied approximately linearly with undeformed chip thickness. However, it should be noted that tests such as those of

Kopalinsky and Oxley (1984) described in section 7.4 show a definite size effect with the friction force increasing less than in proportion to the increase in undeformed chip thickness. A more complete analysis should therefore take account of this.

Neglecting the size effect and noting that α_n and i are both assumed to be zero it follows that for a given work material the friction force per unit undeformed chip thickness along a unit length of the cutting edge has a constant value. Denoting this constant value (friction force intensity) by u, the magnitude of the friction force $|dF|$ acting on a small chip element of which the area of the corresponding undeformed section is dA can be expressed by the relation

$$|dF| = u \, dA = ut(s) \, ds \tag{10.5}$$

where ds is the width of the element and $t(s)$, which varies with the position of the element, is the corresponding undeformed chip thickness. dF itself is a vector quantity. From assumptions (i) and (iii) it acts in general on the chip element in the direction inclined to the undeformed chip thickness vector at an angle equal to the tool local inclination, $i(s)$. A typical differential friction force vector is illustrated in Fig. 10.2 in which an undeformed chip section bounded by a cutting edge of arbitrary

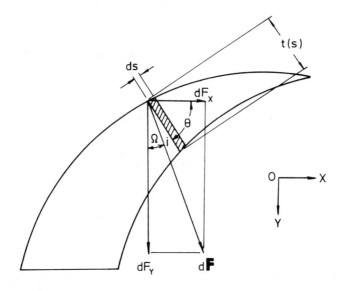

Fig. 10.2 — Resolution of differential friction force at nose radius edge into components in a set of reference directions at right angles: note the X and Y axes lie in the cutting face plane.

shape is shown. In the diagram it is also shown that the undeformed chip element (perpendicular to ds) makes an angle θ with the positive X axis. The angle is defined by the shape of the cutting edge. Therefore the angle Ω that dF makes with the positive Y axis is given by

Sec. 10.2] Prediction of chip flow direction

$$\Omega(s) = \frac{\pi}{2} - i(s) - \theta(s) \tag{10.6}$$

which taking $i = 0°$ reduces to

$$\Omega(s) = \frac{\pi}{2} - \theta(s) \tag{10.7}$$

Using the above expression to add up all vectors by integration gives the resultant friction force

$$F = F_X\hat{i} + F_Y\hat{j} = (\int dF_X)\hat{i} + (\int dF_Y)\hat{j} = (\int |dF| \sin \Omega)\hat{i} + (\int |dF| \cos \Omega)\hat{j} \tag{10.8}$$

where $\hat{i}$ and $\hat{j}$ are unit vectors along the positive X axis and Y axis respectively. On substituting for $|dF|$ from equation (10.5),

$$F = (u\int \sin \Omega \, dA)\hat{i} + (u\int \cos \Omega \, dA)\hat{j} \tag{10.9}$$

where the integral is to be taken over the entire area of the undeformed chip section. As the chip velocity is assumed to be coincident with the friction force vector, the direction of the resultant friction force, namely the chip flow angle $\overline{\Omega}$, measured from the positive Y axis, can be determined from the relation

$$\overline{\Omega} = \tan^{-1}\left(\frac{F_X}{F_Y}\right) \tag{10.10}$$

which on substituting for F_X and F_Y from equation (10.9) gives

$$\overline{\Omega} = \tan^{-1}\left(\frac{\int \sin \Omega \, dA}{\int \cos \Omega \, dA}\right). \tag{10.11}$$

This is the fundamental equation which is used to derive detailed expressions for the predicted chip flow angle for a cut with nose radius tools. It can be seen that the force intensity u is eliminated in forming the ratio F_X/F_Y, and thus the chip flow direction can be predicted in the absence of constants determined from experiments.

Before integrating equation (10.11) it is necessary to express the varying undeformed chip thickness $t(s)$ as a function of chip element position along the active cutting edge. Fig. 10.3(a) shows a typical nose radius tool when the depth of cut is such as to use only the round nose or part of the nose radius edge. The uncut chip section is enclosed by two circular arcs together with a straight line representing the outer work surface. Since the undeformed chip thickness is defined as the length of

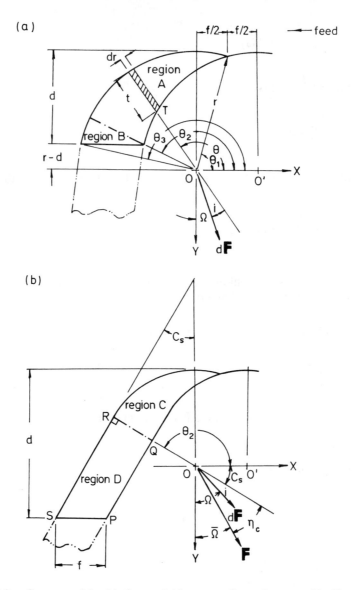

Fig. 10.3 — Geometry of the chip flow model for nose radius tools proposed by Young *et al.* (1987).

the segment intercepted between the chip section boundaries, two expressions are required for region A and region B. In region A, if the thickness segment intersects the inner arc at the point T, the undeformed chip thickness is equal to $(r-\overline{TO})$, where r is the nose radius. Applying the cosine law to the triangle OTO′, it follows that

$$(\overline{TO'})^2 = (\overline{TO})^2 + (\overline{OO'})^2 - 2(\overline{TO})(\overline{OO'})\cos\theta \qquad (10.12)$$

where θ is the angular position of the thickness segment measured counterclockwise from the positive X axis. In fact, $\overline{OO'}$ equals f, the feed, and $\overline{TO'}$ equals r. Substituting these into equation (10.12) and solving the quadratic equation gives

$$\overline{TO} = f \cos \theta + (r^2 - f^2 \sin^2\theta)^{1/2} \qquad (10.13)$$

Consequently

$$t(\theta) = r - f \cos \theta - (r^2 - f^2 \sin^2\theta)^{1/2} \qquad (10.14)$$

for, $\theta_1 \leq \theta \leq \theta_2$, where

$$\theta_1 = \cos^{-1}\left(\frac{f}{2r}\right)$$

and

$$\theta_2 = \pi - \tan^{-1}\left[\frac{r-d}{(2rd-d^2)^{1/2} - f}\right]$$

In region B the undeformed chip thickness becomes

$$t(\theta) = r - (r-d) \operatorname{cosec}(\pi - \theta) = r - (r-d) \operatorname{cosec} \theta \qquad (10.15)$$

for $\theta_2 \leq \theta \leq \theta_3$, where

$$\theta_3 = \pi - \sin^{-1}\left(\frac{r-d}{r}\right)$$

In both situations the elemental friction force is acting in the direction

$$\Omega(\theta) = \theta - \frac{\pi}{2} \qquad (10.16)$$

and is in line with the thickness vector, while the associated diagrams (Fig. 10.3(a,b)) show the force vector for the general case where the inclination angle i is other than zero.

When the cut extends beyond the tool nose to include at least part of the straight side cutting edge, the uncut chip section again needs to be divided into two regions, C and D, as shown in Fig. 10.3(b). The expression for the undeformed chip thickness for region C is the same as for region A of Fig. 10.3(a), namely equation (10.14) applies. However, the upper limit of the boundary, θ_2, should be replaced by $\pi - C_s$. Region D is made up of small thickness segments having the same orientation with the corresponding friction force components acting in the same direction given by

$$\Omega_d = \frac{\pi}{2} - C_s \tag{10.17}$$

Because of this constant value, the factors containing Ω_d can be taken out from the integrals when evaluating equation (10.11) for region D. What remains to be determined is the area of region D, A_d; the expression for the undeformed chip thickness is not required although it varies. The area A_d can be expressed from the associated cut geometry as

$$A_d = \tfrac{1}{2}(\overline{PQ} + \overline{RS})\ \overline{QR} = f[d - r(1 - \sin C_s) - \tfrac{1}{4}f \sin(2C_s)] \tag{10.18}$$

By performing the integrations of equation (10.11) incorporating the above relations, the predicted chip flow angle with respect to the positive Y axis may be written as follows:

(1) Case I (Fig. 10.3(a)), $d < r(1-\sin C_s)$

$$\overline{\Omega} = \tan^{-1}\left[\frac{(\int \sin \Omega\ dA)_A + (\int \sin \Omega\ dA)_B}{(\int \cos \Omega\ dA)_A + (\int \cos \Omega\ dA)_B}\right] \tag{10.19}$$

Substituting dA by $t\ dr$ or $rt\ d\theta$, where dr is the differential of arc, gives

$$\overline{\Omega} = \tan^{-1}\left[\frac{(\int \sin \Omega\ t d\theta)_A + (\int \sin \Omega\ t\ d\theta)_B}{(\int \cos \Omega\ t\ d\theta)_A + (\int \cos \Omega\ t\ d\theta)_B}\right] \tag{10.20}$$

Consequently, by inserting equations (10.14), (10.15) and (10.16), the results for the numerator (NUM) and denominator (DEN) of the equation are given by

$$\text{NUM} = [-r \sin \theta]_{\theta_1}^{\theta_3} + \tfrac{1}{2}\left[\sin \theta\ (r^2 - f^2 \sin^2\theta)^{1/2} + \frac{r^2}{f} \sin^{-1}\left(\frac{f}{r}\sin \theta\right)\right]_{\theta_1}^{\theta_2} +$$

$$+ f\left[\frac{\sin(2\theta)}{4} + \frac{\theta}{2}\right]_{\theta_1}^{\theta_2} + [(r-d)\ ln(\sin \theta)]_{\theta_2}^{\theta_3} \tag{10.21}$$

Sec. 10.2]   Prediction of chip flow direction

$$DEN = [-r\cos\theta]_{\theta_1}^{\theta_3} + \tfrac{1}{2}\bigg[\cos\theta\,(r^2-f^2\sin^2\theta)^{1/2} + \frac{r^2-f^2}{f}\ln\{(f\cos\theta) +$$

$$+ (r^2-f^2\sin^2\theta)^{1/2}\}\bigg]_{\theta_1}^{\theta_2} + \tfrac{f}{4}[\cos(2\theta)]_{\theta_1}^{\theta_2} + [-(r-d)\theta]_{\theta_2}^{\theta_3}$$

where the limits of integration are

$$\theta_1 = \cos^{-1}\!\left(\frac{f}{2r}\right)$$

$$\theta_2 = \pi - \tan^{-1}\!\left[\frac{r-d}{(2rd-d^2)^{1/2}-f}\right]$$

$$\theta_3 = \pi - \sin^{-1}\!\left(\frac{r-d}{r}\right)$$

(2) Case II (Fig. 10.3(b)), $d \geq r(1-\sin C_s)$
Applying the same procedure as for case I gives

$$\bar{\Omega} = \tan^{-1}\!\left[\frac{(\int \sin\Omega\,dA)_C + (\int \sin\Omega\,dA)_D}{(\int \cos\Omega\,dA)_C + (\int \cos\Omega\,dA)_D}\right] \qquad (10.22)$$

Since the angle Ω_d is constant over region D, the equation can be simplified to give

$$\bar{\Omega} = \tan^{-1}\!\left[\frac{(\int \sin\Omega\,dA)_C + \sin\Omega_d\,A_d}{(\int \cos\Omega\,dA)_C + \cos\Omega_d\,A_d}\right] \qquad (10.23)$$

To obtain an explicit expression, equations (10.14) and (10.16) are substituted and integrated over region C. With the aid of equations (10.17) and (10.18), the numerator and denominator of the above equation can be found from the expressions

$$NUM = [-r^2\sin\theta]_{\theta_1}^{\theta_2} + \frac{r}{2}\bigg[\sin\theta\,(r^2-f^2\sin^2\theta)^{1/2} + \frac{r^2}{f}\sin^{-1}\!\left(\frac{f}{r}\sin\theta\right)\bigg]_{\theta_1}^{\theta_2} +$$

$$+ f\bigg[\frac{\sin(2\theta)}{4} + \frac{\theta}{2}\bigg]_{\theta_1}^{\theta_2} + \bigg[f\{d - r(1-\sin C_s)\} - \frac{f^2}{4}\sin(2C_s)\bigg]\cos C_s$$

$$(10.24)$$

$$\text{DEN} = \left[-r^2 \cos\theta\right]_{\theta_1}^{\theta_2} + \frac{r}{2}\left[\cos\theta\,(r^2 - f^2 \sin^2\theta)^{1/2} + \frac{r^2 - f^2}{f}\ln\{(f\cos\theta) + (r^2 - f^2 \sin^2\theta)^{1/2}\}\right]_{\theta_1}^{\theta_2} + \frac{rf}{4}[\cos(2\theta)]_{\theta_1}^{\theta_2} + \left[f\{d - r(1 - \sin C_s)\} - \frac{f^2}{4}\sin(2C_s)\right]\sin C_s$$

where the corresponding limits of integration are

$$\theta_1 = \cos^{-1}\left(\frac{f}{2r}\right)$$

$$\theta_2 = \pi - C_s$$

In view of the above results, the predicted chip flow direction for any depth of cut can be written in the general form as

$$\overline{\Omega} = \tan^{-1}\left(\frac{\text{NUM}}{\text{DEN}}\right) \tag{10.25}$$

where the coefficients NUM and DEN are functions of cut geometry and are found from equations (10.21) and (10.24). In experiments it is more convenient (and accurate) to measure the chip flow angle with reference to the normal to the straight side cutting edge of the tool. If this angle is denoted η_c as before, it can be related to $\overline{\Omega}$, as referred to Fig. 10.3(b), by the relationship

$$\eta_c = \frac{\pi}{2} - C_s - \overline{\Omega} \tag{10.26}$$

It should be noted that the cut geometry considered above excludes the situations where cutting occurs at the straight part of the end cutting edge. These situations are rarely found in machining with nose radius tools in the practical range of cutting conditions. They take place only when the feed is greater than or equal to the nose radius for a tool having a zero side cutting edge angle. Also, the greater the side cutting edge angle is, the greater is the feed required for the cut to include the straight part of the end cutting edge. For instance, when a triangular 'throwaway' tip of 0.4 mm nose radius (the smallest size available in the commercial range) is used on a toolholder of zero side cutting edge angle the corresponding transitional feed is 0.4 mm/rev. On increasing the side cutting edge angle to 30°, the transitional feed increases to 0.7 mm/rev. In fact, a similar formula to equation (10.25) will result if a combination of ultrahigh feed and small nose radius should occur, which causes the cut to involve the straight part of the end cutting edge.

The effect of the various parameters on the predicted chip flow angle is not

Sec. 10.2]   **Prediction of chip flow direction**   171

readily seen from the above equations. In view of this Young *et al.* (1987) made a numerical study in which η_c was calculated from equations (10.25) and (10.26) with the aid of equations (10.21) and (10.24). The conditions (depth of cut, etc.) used in the calculations covered the same range as in their experiments (see next section). The results obtained are represented by the curves in Figs. 10.4, 10.5 and 10.6. As

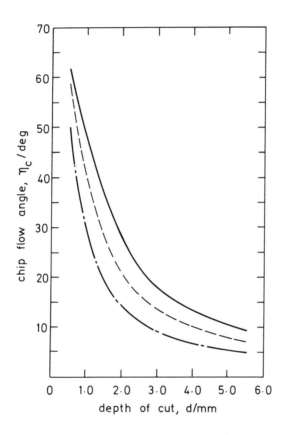

Fig. 10.4 — Effect of depth of cut and nose radius on the predicted chip flow angle: $f = 0.20$ mm/rev; $C_s = 0°$; ———, $r = 1.6$ mm; – – –, $r = 1.2$ mm; —·—, $r = 0.8$ mm.

would be expected all of the curves show similar trends with the predicted chip flow angle increasing with decrease in depth of cut. It can be seen that the chip flow angle rises very rapidly from the point where the depth of cut is about twice the nose radius. The curves show that the chip flow angle has a large value, even for a small radius of 0.8 mm. Also it is of interest to see from the curve for $r = 0.8$ mm in Fig. 10.4 that the chip flow angle does not approach zero as the depth of cut is increased up to 5.6 mm (seven times the nose radius). All these results clearly indicate that the nose radius has a significant influence on the chip flow direction and cannot be ignored in predicting the chip flow angle and, consequently, the cutting forces.

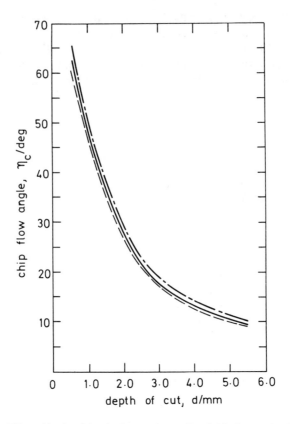

Fig. 10.5—Effect of feed and depth of cut on the predicted chip flow angle: $C_s = 0°$; $r = 1.6$ mm; —·—, $f = 0.40$ mm/rev; ———, $f = 0.20$ mm/rev; – – –, $f = 0.10$ mm/rev.

From the families of curves presented in Figs. 10.4, 10.5 and 10.6 it is clear that the chip flow angle increases with an increase of the nose radius and feed and with a decrease of the side cutting edge angle. The degree of influence of these three parameters can be explained by considering in each case the effect of changing the parameters on the proportion of the engaged nose radius part of the cutting edge to the entire active cutting edge. It can be seen that, as the nose radius is increased with all other conditions being equal, the corresponding radius edge engaged in cutting increases approximately by the same multiple. On the other hand, increasing the feed has a reduced effect, adding only a small segment of radius edge, the length of which is approximately equal to the feed increment, to the original edge. Considering the influence of the side cutting edge angle C_s, the cut geometry shows that the engaged radius edge is proportional to 1.0-sin C_s. Because of the difference in the nature of the variations of the engaged radius edge with the separate parameters, the relative effects of each parameter on the predicted chip flow angle can be expected to be quite different. These relative effects can be examined from the curves in Figs. 10.4, 10.5 and 10.6, where the curves represented by solid lines are formed under the

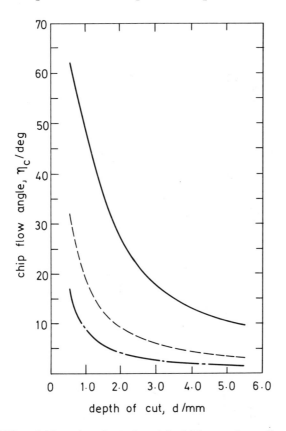

Fig. 10.6 — Effect of side cutting edge angle and depth of cut on the predicted chip flow angle: $f = 0.20$ mm/rev; $r = 1.6$ mm; ———, $C_s = 0°$; – – –, $C_s = 30°$; —·—, $C_s = 45°$.

same cutting conditions and can be used as a reference for making comparison. It can be seen that the effects of the nose radius and side cutting edge angle are somewhat larger than that of the feed, as is expected from the above reasoning. Moreover, when the chip flow angle is plotted directly against the nose radius and side cutting edge angle separately, the former (Figs. 10.7(a,b)) gives approximately a linear relationship while the latter (Fig. 10.8(a,b)) gives a set of concave-upwards curves. This coincides with the relationships of the length of engaged radius edge with the respective parameters discussed above.

10.3 EXPERIMENTAL INVESTIGATION OF CHIP FLOW DIRECTION AND COMPARISON WITH PREDICTED RESULTS

Similar experiments to those described in section 9.2 were made by Young *et al.* (1987) with the chip flow angle measured from still photographs taken of the chip flowing over the tool as shown in Fig. 9.1. The three forces P_1, P_2 and P_3 (Fig. 8.2) were also measured. Since the main purpose of the experiments was to investigate

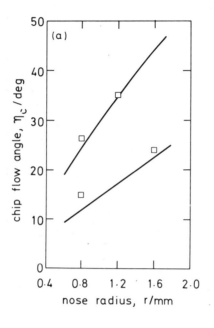
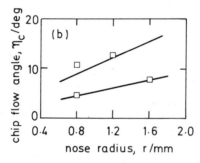

Fig. 10.7 — Predicted and experimental chip flow angles: top line, $d = 1.2$ mm; bottom line, $d = 2.4$ mm: (a) $f = 0.20$ mm/rev; $C_s = 0°$; (b) $f = 0.20$ mm/rev; $C_s = 30°$.

their proposed chip flow model for machining processes with nose radius tools, Young *et al.* placed emphasis on those parameters which were predicted to have most influence on the chip flow angle. In the formulation of the tests an appropriate number of levels were chosen for each selected parameter as now described.

(a) Cutting speed
The cutting conditions were selected to avoid built-up edge. From this consideration the choice of the cutting speeds were in arithmetic progression at the levels of 150, 200 and 250 m/min, which fall in the range of cutting speeds encountered in practical turning operations.

Sec. 10.3] Experimental investigation of chip flow direction 175

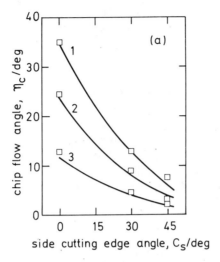

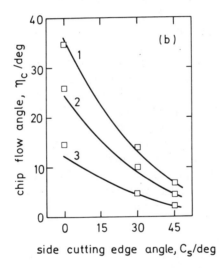

Fig. 10.8 — Predicted and experimental chip flow angles: (a) $f = 0.20$ mm/rev; $r = 1.2$ mm; line 1, $d = 1.2$ mm; line 2, $d = 1.8$ mm; line 3, $d = 3.6$ mm; (b) $f = 0.20$ mm/rev; $r = 0.80$ mm; line 1, $d = 0.80$ mm; line 2, $d = 1.20$ mm; line 3, $d = 2.40$ mm.

(b) Feed

The same considerations as for cutting speed led to the choice of an arithmetic progression for the feed. The levels chosen were 0.1 and 0.2 mm/rev.

(c) Tip nose radius

It is this parameter that forms the focus of the analysis. As discussed earlier in deriving the equations for the chip flow angle, it has been assumed that the nose radius used was equal to or greater than the feed. With this in mind and considering the aim of using practically relevant values while the tests still gave a clear indication of the effects of a change in this parameter on the chip flow angle, three levels of nose radii in arithmetic progression were selected. These were 2/64 in (0.8 mm), 3/64 in (1.2 mm) and 4/64 in (1.6 mm).

(d) Depth of cut
The effect of nose radius on the measured variables depends on the relative value of nose radius to depth of cut. On account of this, a new parameter defined as the ratio of nose radius to depth of cut (r/d) was used in formulating the test conditions. This parameter was varied from 1/3 to 1 in arithmetic progression at three levels, which resulted in the choice of the following three sets of depth of cut with each set corresponding to a selected nose radius: (1) 0.8, 1.2, 2.4 mm for the nose radius $r = 0.8$ mm, (2) 1.2, 1.8, 3.6 mm for $r = 1.2$ mm and (3) 1.6, 2.4, 4.8 mm for $r = 1.6$ mm.

(e) Side cutting edge angle
The proportion of the nose radius edge engaged in cutting is affected by the side cutting edge angle. The larger this angle is, the smaller the contribution of the radius part to the active cutting edge. In the experiments tools with side cutting edge angles of 0°, 30° and 45° were employed.

The work material used in the tests was a steel of chemical composition 0.18% C, 0.04% Si, 1.40% Mn, 0.04% P, 0.04% S. The cutting tools used were 'throwaway' tips which locked on specially designed toolholders which had neutral (0°) rake and inclination angles and side cutting edge angles of 0°, 30° and 45°. The tips chosen were triangular carbide tips (without a chip control groove) with an ISO grading of P10. These were specially ground and lapped to give a small clearance angle of 5° on the side of the tip while taking care to maintain the nose radius. After this the outlines of each tip were checked carefully under a Nikon shadowgraph with master profiles drawn on paper; the defective edges were then marked and kept aside and not used. Finally each qualified corner of the tips was numbered and randomly assigned to a test.

Noting that van Luttervelt and Pekelharing (1976) and Kluft *et al.* (1979) had indicated that the curvature of the workpiece influenced the chip flow direction Young *et al.* made preliminary tests to check this. These were made on two short cylindrical tubes machined to outside diameters of 60 and 120 mm with the tool edge set normal to the cutting velocity. Both tubes had the same wall thickness of 4.8 mm. The tests were carried out under orthogonal machining conditions at a constant depth of cut (4.8 mm) while allowing the feed and cutting speed to vary. The levels of feed and cutting speed chosen were the same as given above. The results of the tests showed that the chip flow direction was invariably (approximately) perpendicular to the cutting edge. Thus the choice of the workpiece diameter for the conditions considered was immaterial. In order to reduce the variation of cutting velocity across the cutting edge, however (to err on the side of safety), large workpiece diameters of approximately 140 mm diameter were used. In this case, a variation of less than 2% in the cutting velocity from the mean was introduced.

Experimental values of chip flow angle η_c (Fig. 10.3(b)) together with predicted values for the same conditions are given in Figs. 10.7 and 10.8. As mentioned before

η_c was measured from the straight part of the cutting edge as shown in Fig. 10.3(b). In measuring η_c from the photographs care was taken to make the measurements as near to the cutting edge as possible. If the chips were not straight but curved, the tangent to the chip edge at the intersection with the cutting edge was taken as the chip flow direction. A small number of such cases were found where the depths of cut were small (for example $d = 0.8$ and 1.2 mm). Replication tests were made if there was evidence of interference between the removed chip and the surrounding objects. Because there was little change in η_c with cutting speed, as observed from the experimental results, the experimental values of η_c were taken to be the average values for the three cutting speeds used. The agreement between predicted and experimental values of η_c can be seen from Figs. 10.7 and 10.8 to be very good with the experimental results, confirming the predicted trends. Two other methods for predicting η_c for nose radius tools which like the method of Young et al. do not include empirical factors have been proposed by Colwell (1954) and Okushima and Minato (1959) as described earlier. Predictions made from the relevant equations, namely equations (10.1), (10.2) and (10.3) for Colwell and equation (10.4) for Okushima and Minato together with those from the model of Young et al. are given in Figs. 10.9(a–c). The results are for conditions typical of those used by Young et al. in their experiments. It can be seen that all three predicted curves follow the experimental trends very well. However, the methods suggested by Colwell and by Okushima and Minato overestimate η_c while the method of Young et al. gives excellent agreement — less than 15% deviation from the experimental results.

10.4 PREDICTION OF CUTTING FORCES AND COMPARISON WITH EXPERIMENTAL RESULTS

Having shown that the chip flow direction could be predicted accurately for nose radius tools Young et al. (1987) again introduced the concept of an equivalent cutting edge in order to predict forces. In this analysis the side cutting edge angle of the equivalent cutting edge is taken to be

$$C_s^* = C_s + \eta_c \tag{10.27}$$

so that for the simplified conditions considered ($i = \alpha_n = 0°$) the chip flow direction will be normal to the equivalent cutting edge, which seems a reasonable assumption. The nose radius tool is therefore treated as if it was a tool having a single straight cutting edge with the width of cut given by $w = d/\cos C_s^*$ and the undeformed chip thickness by $t_1 = f \cos C_s^*$. The cutting force calculations are somewhat simpler than those described in the previous chapter because, for the conditions assumed ($i = 0°$), $F_R = 0$ and P_1, P_2 and P_3 are found from equations (8.16) simply in terms of F_C and F_T with $C_s = C_s^*$. In the calculations the temperature factors η and ψ were both taken as 0.7. The work material properties used were those appropriate to the 0.18% carbon steel used in the experiments. Predicted force results obtained in this way are represented by the lines in Figs. 10.10 to 10.13.

It can be seen from Figs. 10.10–10.13, which also give the experimental force results obtained by Young et al., that for all of the conditions considered there is good

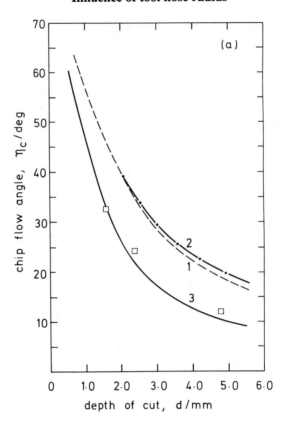
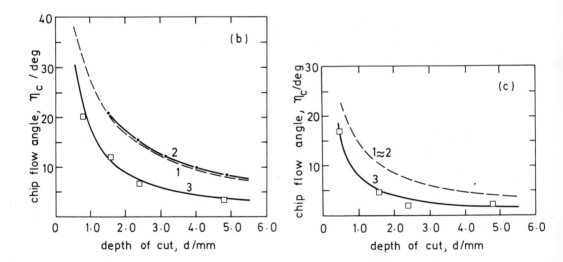

Fig. 10.9 — Comparison of experimental chip flow angles with chip flow angles predicted by different methods: $f = 0.10$ mm/rev; $r = 1.6$ mm; line 1, predicted results based on Colwell's method; line 2, predicted results based on Okoshima and Minato's method; line 3, predicted results based on Young et al.'s method: (a) $C_s = 0°$; (b) $C_s = 30°$; (c) $C_s = 45°$.

Sec. 10.4] Prediction of cutting forces

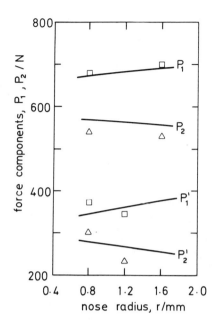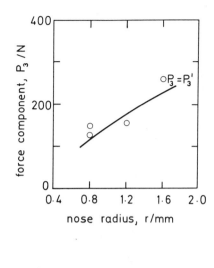

Fig. 10.10 — Predicted and experimental cutting forces: $U = 150$ m/min; $f = 0.1$ mm/rev; $C_s = 0°$; P_1, P_2, P_3 are forces for $d = 2.4$ mm; $P'_1 P'_2 P'_3$ are forces for $d = 1.2$ mm.

agreement between predicted and experimental results. Variations of cutting forces with nose radius are given in Fig. 10.10. These results show that with all other cutting conditions held constant an increase in nose radius has little effect on P_1 and P_2 but gives a considerable increase in P_3. The depth of cut can be seen to have little if any influence on P_3. This can also be seen from the results in Fig. 10.11 which shows the influence of depth of cut and feed on forces. The results in Fig. 10.12 show that all of the cutting forces are predicted to decrease with increase in cutting speed and the experimental results confirm this. The mechanism by which an increase in speed decreases the chip thickness and hence decreases forces has been explained in section 7.4 in terms of the machining theory. Fig. 10.13 shows the influence of side cutting edge angle on forces. These results indicate that an increase in side cutting edge angle gives the same trends as an increase in nose radius (Fig. 10.10), namely, a gradual increase and decrease in P_1 and P_2 respectively but a relatively rapid increase in P_3. This can be explained from the equivalent cutting edge model. As seen before, the chip flow angle increases with increase in nose radius. Consequently, when the actual cutting edge is replaced by an equivalent cutting edge, which is normal to the predicted chip flow direction, the net effect of increasing the nose radius is to increase the side cutting edge angle.

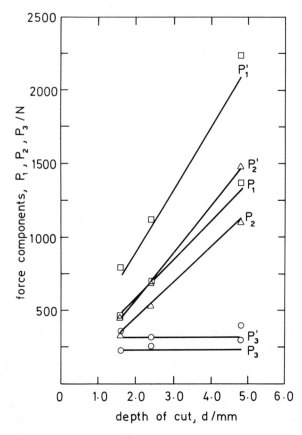

Fig. 10.11 — Predicted and experimental cutting forces: $U = 150$ m/min; $C_s = 0°$; $r = 1.6$ mm; P_1, P_2, P_3 are forces for $f = 0.1$ mm/rev; $P'_1 P'_2 P'_3$ are forces for $f = 0.2$ mm/rev.

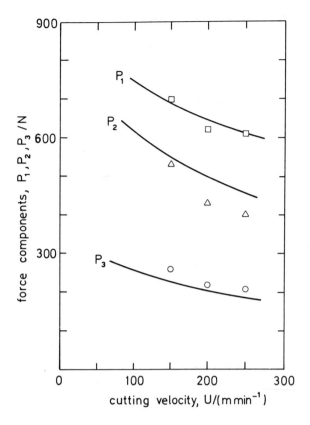

Fig. 10.12 — Predicted and experimental cutting forces: $f = 0.1$ mm/rev; $C_s = 0°$; $r = 1.6$ mm; $d = 2.4$ mm.

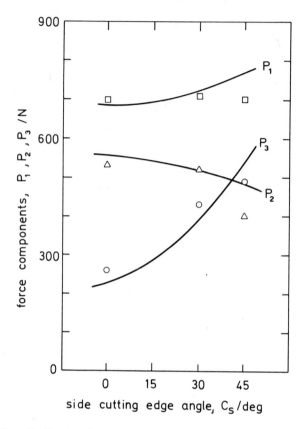

Fig. 10.13 — Predicted and experimental cutting forces: $U = 150$ m/min; $f = 0.1$ mm/rev; $r = 1.6$ mm; $d = 2.4$ mm.

11

Assessing machinability factors in terms of machining theory

11.1 INTRODUCTION

In this chapter it is shown how the machining theory developed in this book can be applied to relate such machinability factors as cutting power, tool life and surface finish to work and tool material properties and cutting conditions. These relations are needed at the simplest level for such tasks as grading the machinability of work materials. At a more sophisticated level they are needed in selecting optimum cutting conditions based on criteria such as minimum cost or maximum production rate and subject to constraints such as cutting power available and required surface finish on produced components — see Shaw (1984) for a discussion of machining process optimisation. As mentioned in section 1.2 it has been usual to date to determine machinability relationships from data collected from machining tests and a vast amount of work has been carried out to this end. The approach based on the machining theory now described offers the possibility of greatly reducing this work. It is known that other factors such as the design and condition of machine tools can affect tool life, surface finish, etc. However, this is outside the scope of the present treatment of the problem and the role of the machine tool from the viewpoint of effects such as regenerative chatter is ignored. A further limitation is that because no direct account has been taken of cutting fluids in developing the machining theory little consideration can be given to their influence on processes.

11.2 CUTTING FORCES AND POWER

It has been shown in the preceding chapters that cutting forces can be predicted with reasonable accuracy from the machining theory for a wide range of plain carbon steel work materials with the influence of speed, feed and cutting edge geometry accurately accounted for. It is therefore possible to make good estimates of the cutting power which for the turning operations considered is given by the product of the force in the cutting direction and the cutting speed. The limiting factor in determining the rate of metal removal is often the machine tool power available and the accurate prediction of the cutting power is therefore extremely important in this sense. The accurate prediction of forces is also important in considering the elastic deflections of the machine tool which can affect the accuracy of the machined component. Also, if the analysis is at some stage to be extended to consider the stability of the machine tool then it will be necessary to know how forces vary under dynamic conditions in which the speed, feed and cutting edge geometry can vary as a result of vibrations in the system. Such data have usually been obtained experimentally (see Tobias 1965) and it would clearly be an advantage if machining theory can be used to generate the required relationships.

11.3 TOOL WEAR AND TOOL LIFE

In a well-designed machining operation the useful life of the cutting tool will normally be determined by the wear of the cutting edge (or edges). Wear can occur on the cutting face and clearance face of the tool as shown diagrammatically in Fig. 11.1. The wear on the cutting face is in the form of a small depression called a crater

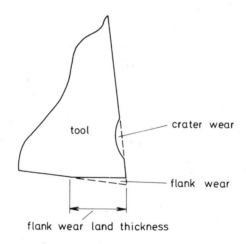

Fig. 11.1 — Diagrammatic representation of tool cutting edge wear.

situated just back from the cutting edge and the dimensions of this crater may be used to define the end of effective tool life. Wear on the clearance face, usually referred to as flank wear, appears as a land immediately below the cutting edge and the average (or maximum) thickness of this land may be used to define the end of effective tool life. Since for many conditions little crater wear occurs it is usual in practice to define tool life in terms of flank wear. This is also physically more meaningful in that the surface finish on the machined component is usually related to the degree and form of the flank wear and can become unacceptable at a certain level of wear.

Sec. 11.3] **Tool wear and tool life** 185

The empirical tool life equation introduced by Taylor (1907)

$$LU^a f^b = C \tag{11.1}$$

where L is the tool life (based on flank or crater wear or total tool failure), U is the cutting speed and f is the feed with a, b and C constants, is still widely applied in industry and is the basis of most present-day optimisation procedures. Unfortunately it does not agree well with experimental results over wide ranges of cutting conditions, giving best agreement over narrow ranges of cutting speed and feed. This is not surprising since the plot of log L versus log U is quite often curved with possible turning points at slow speeds which means that the speed index a will vary with the cutting conditions selected for the tests. Further, the indices a and b are not independent although often treated as such. It is understandable therefore that considerable effort has been devoted to the search for more fundamental tool life relationships. A less empirical approach is suggested by work on the basic mechanisms of tool wear such as that of Trent (1977), Loladze (1962) and Takeyama and Murata (1963) which shows that when machining with cutting tool temperatures in excess of 800°C, which would generally be the case for carbide tools in the normal cutting speed range, the main wear mechanism is diffusion which is a temperature-controlled rate process. Therefore, if the appropriate temperature can be determined in terms of work material properties and cutting conditions from machining theory and the relationship between tool wear rate (and hence tool life) and temperature is known it should be possible to predict tool life with far less effort than by purely empirical means. This approach is now considered.

Experimental results obtained by Takeyama and Murata (1963) from turning tests showing the influence of tool temperature on tool life (defined in this case as the time for the flank wear land to reach an average thickness of 0.6 mm) are given in Fig. 11.2. In these tests temperatures were measured using a tool–work thermocouple technique. The work material was a heat-resistant steel and the tool material a P10 grade of carbide. A single non-orthogonal tool geometry was used. The cutting speed was varied from 5 to 270 m/min and the feed from 0.05 to 0.25 mm/rev. It can be seen (Fig. 11.2) that when plotted on a log–log basis all of the experimental results closely follow a single straight line, thus indicating a relationship of the form

$$L = AT_{\text{tool}}^{-B} \tag{11.2}$$

where L is the tool life, T_{tool} is the tool temperature and A and B are constants. Woxen (1937) and Colding (1959) have also shown good agreement between flank wear results and equation (11.2) while Trigger and Chao (1956) and Ling and Saibel (1956) have shown good agreement with equation (11.2) when tool life is defined in terms of cutting face crater wear. All of these workers assumed that diffusion was the dominant cause of wear. It should be noted that for temperatures lower than 800°C the results of Takeyama and Murata (1963) show that the flank wear becomes independent of temperature and dependent on the sliding distance as in abrasive wear. In applying the method of predicting tool life now described which is based on

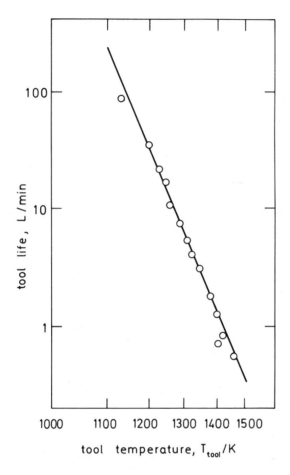

Fig. 11.2 — Experimental tool life and tool temperature results.

equation (11.2) care should therefore be taken to ensure that the temperatures are sufficiently high for diffusion to be the predominant wear mechanism.

In a co-operative research programme organised by OECD and CIRP (1966) which was carried out in laboratories all over the world experimental results including cutting forces, temperatures and tool life were obtained from turning tests for a steel work material of chemical composition 0.44% C, 0.41% Si, 0.71% Mn, 0.01% S, 0.015% P, 0.09% Ni, 0.08% Cr, 0.046% Cu. Fortunately, some of the tool life results were obtained using a P10 grade of carbide (results were also obtained for a P30 carbide) which is the same tool material as used by Takeyama and Murata (1963) in obtaining the results in Fig. 11.2. Noting this Oxley et al. (1974) used Takeyama and Murata's tool life results, which can be represented by the relation

$$L = 10^{71.75} T_{\text{tool}}^{-22.82} \tag{11.3}$$

where L is the tool life in minutes for 0.6 mm flank wear and T_{tool} is the temperature in kelvins, together with temperatures determined from the machining theory, for the conditions in the OECD/CIRP experiments, to predict tool life and compared their results with the corresponding experimental tool life results. This approach can be seen to neglect differences in the work materials in the two sets of tests other than in their effect on temperature. Some support for this is given by the work of Shallbroch and Schaumann (1937) who measured tool life and temperatures experimentally and showed that their results for two different steel work materials followed a single line when tool life was plotted against temperature.

In the OECD/CIRP experiments the side cutting edge angle C_s was not zero in all cases and the tool had a nose radius of 0.8 mm compared with a depth of cut of 3.0 mm. In spite of this Oxley *et al.* assumed that the process could be treated as approximately orthogonal with the undeformed chip thickness and width of cut suitably adjusted for the given value of C_s in the way described in section 8.3. In calculating temperatures an early form of the theory described in Chapter 7 was used with the strain-rate constants taken as $C = 5.9$ and $\delta = 0.05$ and with the tool–chip contact length calculated from equation (6.17) and not from equation (7.13). The temperature factors η and ψ were both taken as unity. The flow stress and thermal properties for the work material were obtained from the equations in Appendix A3 and from equations (6.18), (6.19) and (6.20) and the density ρ taken as 7862 kg/m^3. The tool life in the OECD/CIRP experiments was defined as the time to reach 0.2 mm flank wear and in predicting tool life from equation (11.3), which is for 0.6 mm flank wear, Oxley *et al.* simply assumed that the life would be a third of the value given by this equation. This supposes that the flank wear against time curve is a straight line passing through the origin which in general is an oversimplification of the actual relation. It should, however, be sufficiently accurate for the required purpose. In making tool life predictions from equation (11.3) it was decided that it would be most appropriate to take T_{tool} as equal to the flank temperature T_{flank}. Boothroyd (1963) and Boothroyd *et al.* (1967) have shown from experimental temperature measurements that the ratio of flank to interface temperature (K) varies between 0.82 and 0.95 with an average value of 0.89. T_{tool} was therefore found from the values of T_{int} given by the machining theory from the relation

$$T_{tool} = T_{flank} = 0.89 T_{int} \tag{11.4}$$

where all temperatures are in kelvins.

Predicted tool life results obtained in the way described above are represented by the full line in Fig. 11.3 in which the OECD/CIRP experimental tool life results are also given. The predicted results can be seen to be in good agreement with the experimental results although there is considerable scatter in the experimental results about the predicted line which is typical for tool life test results. Oxley *et al.* (1974) have suggested the following possible explanation for this scatter. In the OECD/CIRP co-operative research programme a series of slow speed compression tests were made on specimens cut from different sections of the bars of work material used in the machining tests. The results of the compression tests when plotted on the basis of equation (4.14) showed that σ_1 varied by approximately $\pm 7\%$ about the

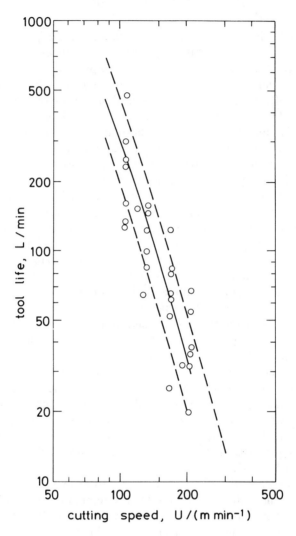

Fig. 11.3 — Predicted and experimental tool life results: ———, predicted results for nominal work material; – – –, predicted results for ±7% variation about the mean values for σ_1.

mean value. Assuming that this variation would still apply at the strain-rates and temperatures encountered in machining, Oxley *et al.* recalculated tool life values in the same way as above but with the values of σ_1 obtained from the equations in Appendix A3 varied by ±7%. The results obtained are represented by the broken lines in Fig. 11.3 and can be seen to encompass 20 of the 28 experimental data points. This lends some support to the often expressed view that hardness ($\sigma_1 \propto$ hardness) variations are one of the main causes of the tool life variations which occur when machining different parts of the same bar or different bars of nominally the same material.

Sec. 11.3] Tool wear and tool life

Following a suggestion by Oxley *et al.* (1974) an alternative method of predicting tool life was developed by Hastings and Oxley (1976). In this, experimental values of tool life were used in conjunction with temperature values determined from the machining theory, as opposed to experimentally measured temperatures, to obtain the constants A and B in equation (11.2). To do this Hastings and Oxley used some experimental tool life results obtained by Bruce (1972) from turning tests on an En8 steel work material using P30 grade carbide tools. Using temperatures obtained from the machining theory for properties appropriate to the En8 steel, in the same way as described above for cutting conditions corresponding to those used in Bruce's tool life tests, they obtained the relation

$$L = 10^{50.38} \, T_{\text{tool}}^{-16.06} \qquad (11.5)$$

where L is the tool life in minutes for 0.2 mm flank wear. Predicted tool life results for the OECD/CIRP work material obtained as before but with equation (11.5) used in place of equation (11.3) are given in Fig. 11.4; the broken lines again represent a variation of $\pm 7\%$ in σ_1. The experimental tool life results in Fig. 11.4 are those obtained in the OECD/CIRP tests for P30 grade carbide tools. The agreement between predicted and experimental results is again good although the predicted lines should clearly be steeper. In this connection it is worth pointing out that equation (11.5) was obtained on the basis of only four experimentally measured values of tool life L and this could well be insufficient. Further work will be needed to clarify this point.

Hastings *et al.* (1979) have used turning tests to measure the rate of tool flank wear for a range of steel work materials, carbide tool materials and cutting conditions. They calculated cutting temperatures from the machining theory in its full form as represented by the chart in Fig. 7.12. In the calculations η was taken as 0.7 but ψ was allowed to vary between 0.7 and 1.0 to take account of the influence of flank wear on temperature. For each combination of work material–tool material considered it was found that the logarithm of the experimental wear rate values followed (approximately) a straight line when plotted against the reciprocal of the corresponding values of estimated cutting temperature, taken as T_{int}, thus indicating a temperature-dependent wear mechanism such as diffusion. For a given tool material and temperature the results showed that the wear rate could vary markedly with work material. The good agreement between predicted and experimental results in Figs. 11.3 and 11.4 could therefore be somewhat fortuitous as the tool life–temperature relationships (equations (11.3) and (11.5)) used in making the predictions were obtained from tests made on work materials different from those used in obtaining the experimental tool life results in Figs. 11.3 and 11.4. It is clear that in predicting tool life the tool life–temperature relationship should preferably be obtained for the work material–tool material combination considered.

Even with the above limitation the method of predicting tool life using equation (11.2) is apparently far more effective than methods based on equation (11.1) and similar equations. This is because the machining theory allows the parameters of speed, feed and rake angle to be combined into the single parameter of temperature. Thus the constants A and B in equation (11.2) can be obtained from a relatively small

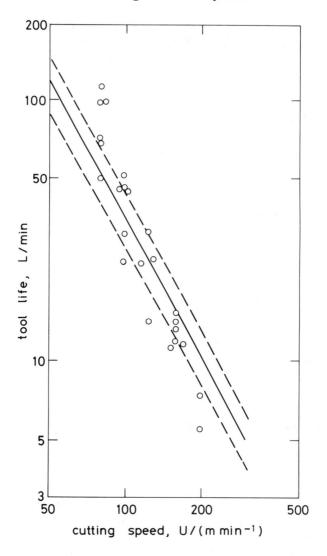

Fig. 11.4 — Predicted and experimental tool life results: ———, predicted results for nominal work material; − − −, predicted results for ±7% variation about the mean values for σ_1.

number of tests and the resulting equation used to predict tool life over a much wider range of conditions. It might also be expected that with this approach allowance can be made for variations in the properties of nominally the same work material without having to redetermine the values of A and B for each variation. The results in Figs. 11.3 and 11.4 which show the influence on tool life of variations in hardness of the work material support this view.

11.4 CUTTING EDGE DEFORMATION

Cook and Goldberger (1982) have pointed out that in machining operations it is usual to select the feed in a rather arbitrary fashion and then to optimise the process on the basis of cutting speed. This approach is surprising when it is considered that from the viewpoints of both specific cutting energy (energy per unit volume of metal removed) and tool life it is normally more beneficial to increase the rate of metal removal by increasing the feed rather than by increasing the speed. With this in mind Cook and Goldberger have outlined a method of selecting the feed to give maximum rate of metal removal and this is demonstrated by the results given in Fig. 11.5, which

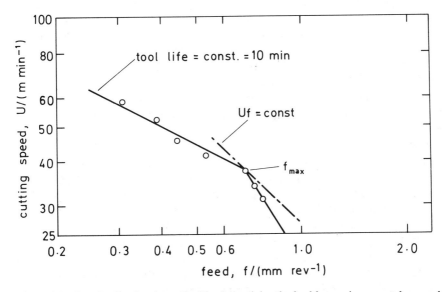

Fig. 11.5 — Results showing the method for determining the feed for maximum metal removal rate.

were obtained from tool life tests when turning a 4340 steel of 330 Brinell hardness using a Carboloy 370 tool. For each feed, tests were made at a number of cutting speeds and the results were cross-plotted to determine the speed which would give a flank wear land of 0.25 mm after 10 min of cutting. When plotted on a log–log basis it can be seen that the results can be represented by two straight lines. On the same basis constant metal removal rate conditions (Uf = constant) can be represented by straight lines having a slope of -1, the further the line being to the right the greater the rate of removal. The rate of removal can therefore in principle be increased by increasing the feed to the point where the slope of the constant tool life curve is equal to -1, or for the results in Fig. 11.5 to the feed value at the intersection of the two straight lines representing the tool life results. Cook and Goldberger have referred to this feed value as f_{max} and have shown how its value decreases as the hardness of the 4340 steel is increased. For all of their tests they found that feeds greater in value than

f_{max} were needed to cause fracture of the cutting edge. Although Cook and Goldberger offer no explanation of why their results show that it is necessary to reduce cutting speed far more rapidly with increase in feed above f_{max} (Fig. 11.5) in order to maintain the same tool life, it appears that the reason could be that f_{max} corresponds to those conditions where the cutting edge starts to deform plastically. This would be consistent with the experimental results of Trent (1977) which show that accelerated rates of flank wear can occur when the clearance face is bulged outwards by the high stresses acting on the cutting face. Indeed, Trent has suggested that the high rates of metal removal attainable with tungsten carbide–cobalt tools results mainly from their ability to resist plastic deformation of the cutting edge at high temperatures.

Cook and Goldberger have concluded that the experimental determination of f_{max} for a wide range of work material/tool material combinations using their experimental method is not a practical proposition because of the vast amount of work which would be involved. Noting this Nachev and Oxley (1985) have considered how tool stresses and temperatures predicted from the machining theory might be used to determine those cutting conditions which give plastic deformation of the cutting edge. In order to do this it is clearly necessary to know the compressive strength at high temperatures of the considered tool material and it is fortunate that Trent (1967) has obtained such data from hot compression tests for a range of tool materials. Nachev and Oxley's investigation is now described.

Lacking the facilities for carrying out Trent's type of tests Nachev and Oxley selected for their investigation a tool material as near as possible to one of those tested by Trent. Typical results obtained by Trent are given in Fig. 11.6 where the compressive strength at different temperatures is represented by the 5% proof stress value. The tool material chosen by Nachev and Oxley was a Seco-Titan grade S25M of chemical composition WC + 6.5% TiC + 9.5% Co + 14.5% (TaC + NbC). This was considered near enough to the alloy of chemical composition WC + 5% TiC + 9% Co tested by Trent to use his curve for this material (Fig. 11.6) to represent the hot strength properties of the S25M tool material although it might be expected that this would be slightly stronger because of its higher TiC content. Two work materials having chemical compositions 0.19% C, 0.27% Si, 0.88% Mn, and 0.48% C, 0.3% Si, 0.80% Mn were used by Nachev and Oxley in their machining tests.

In predicting the conditions which will cause plastic deformation of the cutting edge the factors of interest are the tool stresses and temperatures. The machining theory assumes a perfectly sharp tool with no forces acting on the clearance face and only provides values of average stresses and temperatures acting on the cutting face. Clearly a detailed analysis of cutting edge deformation should consider the stress and temperature distributions around the cutting edge but lacking such information a more approximate approach must suffice. The approach employed by Nachev and Oxley was to assume that the average tool–chip interface temperature T_{int} could be taken to represent the cutting edge temperature and that the average normal stress at the interface σ_N would be the stress which determines whether or not plastic deformation of the cutting edge occurs. Values of σ_N and T_{int} were calculated from the machining theory in the way summarised in Fig. 7.12. The temperature factors η and ψ were both taken as 0.7. The work material flow stress and thermal properties

Sec. 11.4] Cutting edge deformation

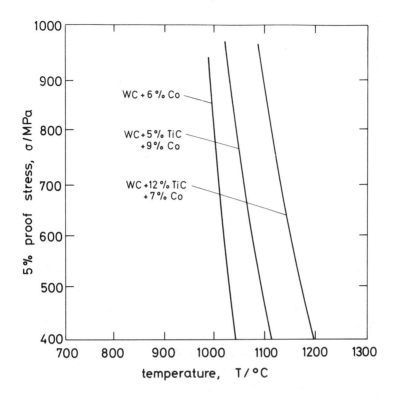

Fig. 11.6 — 5% proof stress values for cemented carbide tool materials obtained from hot compression tests.

were obtained from the equations in Appendix A3 and from equations (6.18), (6.19) and (6.20). The density ρ was taken as 7862 kg/m³. The calculations were made for feeds (undeformed chip thicknesses) in the range from 0.2 to 0.5 mm/rev and cutting speeds in the range from 100 to 500 m/min. Only one rake angle $\alpha = 5°$, which was the rake angle of the tips used in the machining tests, was considered. Although the nose radius of the tips was not negligible compared with the width of cut (0.8 mm compared with 2.5 mm) no account of this was taken in the calculations. The results obtained for σ_N and T_{int} are shown plotted in Fig. 11.7 together with the curve which it is assumed represents the hot compressive strength properties of the S25M tool material used in the machining tests. The σ_N versus T_{int} lines given in Fig. 11.7 are lines of constant feed (undeformed chip thickness) with cutting speed increasing along the lines from left to right. In theory, following the assumptions made, the intersections of the σ_N versus T_{int} curves with the curve representing the compressive strength of the tool material will give the feed–speed combinations at which the tool will start to deform plastically. These values are given in Table 11.1 for the 0.19% carbon steel and in Table 11.2 for the 0.48% carbon steel.

In order to determine the accuracy of their predicted results Nachev and Oxley made machining tests on bars of the 0.19% and 0.48% carbon steels using S25M

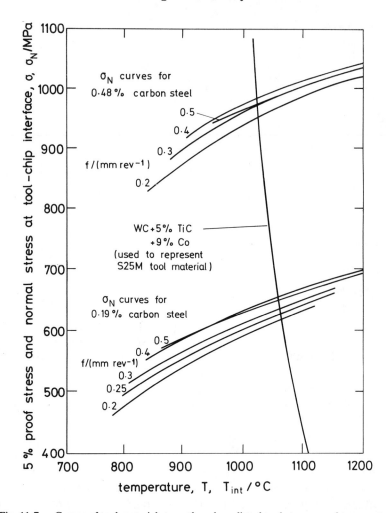

Fig. 11.7 — Curves of tool material strength and predicted tool stresses and temperatures.

Table 11.1

Feed/(mm/rev)	Speed/(m/min)	σ_N/MPa	T_{int}/°C
0.2	450	622	1067
0.25	395	630	1065
0.3	360	642	1064
0.4	290	660	1060
0.5	260	654	1062

Table 11.2

Feed/(mm/rev)	Speed/(m/min)	σ_N/MPa	T_{int}/°C
0.2	250	950	1018
0.25	215	955	1016
0.3	190	960	1012
0.4	170	966	1008
0.5	140	963	1010

grade throwaway tip cutting tools having a positive rake angle of 5°. A conventional turning process was used with the main cutting edge of the tool set normal to the cutting and feed velocities. Although a bar and not a tube was used so that there was some cutting on the end cutting edge (and on the nose radius) the conditions approximated closely to those assumed in the machining theory. The feeds used were the same as those used in making predictions and the starting cutting speed in each test was selected to be approximately the same as the speed at which deformation was predicted to occur (see Tables 11.1 and 11.2). If in this initial test no deformation was observed further tests were made with the speed increased in increments of 50 m/min until deformation was observed, while if deformation was observed further tests were made with the speed decreased in decrements of 50 m/min until no deformation was observed. A new cutting edge was used in each test and the tests were carried out in a random manner to minimise the influence of variations in work and tool material properties, etc. In all tests it was found that 20 s of cutting time was sufficient for steady-state conditions to be achieved. A number of methods were tried for measuring the plastic deformation of the cutting edge and the following method, although the simplest, gave very good results. The clearance face of the tool, which was expected to bulge outwards when plastic deformation occurred, was polished with Brasso to obtain a glossy surface which was then subjected to a strong light and examined with a magnifying glass. If deformation had not occurred the light was reflected from the polished surface as if it was a mirror while if deformation had occurred the light was refracted by the bulged surface and a fringe pattern formed. With this method it was found important to not use too large a magnification as then the surface no longer appears shiny because the structure of the tool and grinding scratches, etc., become visible. Results obtained in this way were later confirmed using a more exact interferometry method.

Experimental results are given in Figs 11.8(a) and 11.8(b) for the 0.19% carbon steel and 0.48% carbon steel respectively with the filled-in symbols indicating the cutting speeds for which plastic deformation was observed and the open symbols indicating cutting speeds 50 m/min lower for which no plastic deformation was observed. Also given are the predicted results from Tables 11.1 and 11.2 which can be represented as straight lines when plotted on a log(speed)–log(feed) basis as in Figs. 11.8(a,b).† Experimental results showing plastic deformation of the cutting

† This method of plotting results is the same as that used by Trent (1977) in his well-known machining charts. In these he has, on the basis of experimental results, indicated the combinations of feed and speed which give plastic deformation of the cutting edge for particular combinations of work and tool materials. The same form of data is also normally given on these charts for crater wear and built-up edge formation.

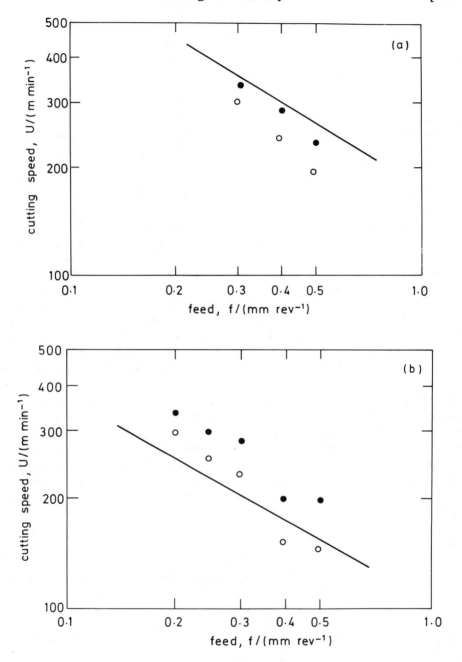

Fig. 11.8 — Predicted and experimental cutting edge plastic deformation conditions: ●, experimental results for which plastic deformation occurred; ○, experimental results for which no plastic deformation occurred: (a) 0.19% carbon steel; (b) 0.48% carbon steel.

edge for the 0.19% carbon steel could not be obtained for feeds of 0.2 mm and 0.25 mm (Fig. 11.8(a)) because the lathe used had insufficient power to reach the required speeds. The agreement between the predicted deformation line and the experimental deformation results can be seen to be very good for the 0.19% carbon steel (Fig. 11.8(a)). It is less good for the 0.48% carbon steel (Fig. 11.8(b)) where the predicted results lie well below the experimental results. The agreement could be improved in this case by moving the curve representing the strength of the S25M tool material in Fig. 11.7 to the right at higher stress values, thus making this curve even steeper than it is. In fact by using a vertical line with $T \approx 1070°C$ to represent the strength of the tool material reasonable agreement could be obtained between predicted and experimental results for both work materials. This would mean that the already small dependence of deformation on stress exhibited by Trent's results had been replaced by essentially no dependence on stress, with temperature becoming the sole criterion of deformation. This would be difficult to believe and other explanations should be considered including the possibility that the machining theory has overestimated the temperatures for the 0.48% carbon steel. If this was so and could be corrected then for a given feed a higher cutting speed would be required to reach the values of σ_N and T_{int} needed to predict deformation (Fig. 11.7) and the agreement between predicted and experimental results in Fig. 11.8(b) would be improved. Further work, both theoretical and experimental, will be necessary to resolve these difficulties. However, even the achieved level of agreement between predicted and experimentally observed deformation conditions should be sufficiently accurate for some practical applications, including making estimates of the maximum metal removal rates which can be achieved without tool deformation's occurring.

11.5 SURFACE FINISH

The best attainable surface finish in a machining operation is determined by such factors as the shape of the cutting edge profile and the feed. For bar turning operations Chisholm (1946, 1949) has obtained expressions for both the centre-line average roughness and peak-to-valley roughness in terms of the tool nose radius, side and end cutting edge angles and the feed. Chisholm (1946) and Ansell and Taylor (1962) have made turning tests which show that this ideal finish is only approached when the cutting conditions are such as to ensure that material is removed in the form of a continuous chip without built-up edge. The presence of a built-up edge or the occurrence of a discontinuous type of chip with cracks running into the newly machined surface can lead to poor surface finish well removed from the ideal. It is therefore important in selecting cutting conditions, particularly for finishing operations, to know how these are related to built-up edge and discontinuous chip formation. It was described in section 7.4 how the machining theory might be used to predict, for plain carbon steel work materials, those cutting conditions giving a built-up edge, or conversely those which do not give a built-up edge, and further consideration is now given to this.

In section 7.4 it was proposed that if the value of T_{mod} at the tool–chip interface was greater than 700 K there would be no built-up edge and even if T_{mod} at the interface was less than 700 K there would be no built-up edge if the temperature at the interface T_{int} was greater than 1000 K. For lower temperatures it was assumed that there would be a built-up edge†. Jahja (1986) has made a detailed study of the effectiveness of these two criteria in predicting the built-up edge range when machining plain carbon steels and this is now described. Bar turning tests were carried out using both tungsten carbide and high speed steel tools on work materials of chemical composition 0.19% C, 0.88% Mn, 0.27% Si, 0.021% P and 0.36% C, 0.89% Mn, 0.10% Si, 0.055% P. Tests were made on the 0.19% carbon steel using rake angles of 5° and − 5° while the tests on the 0.36% carbon steel were limited to a single rake angle of 5°. The conditions in the tests were approximately orthogonal with the undeformed chip thickness equal to the feed. The combinations of feed and cutting speed used were selected to be in the predicted built-up edge range which gave feeds ranging from 0.025 to 0.45 mm/rev and speeds from 20 to 300 m/min. Chips were collected for each test and mounted, polished and etched for examination to determine whether or not a built-up edge was formed during the test. The results are plotted in Fig. 11.9, on the same basis as in a Trent machining chart, with those combinations of feed and speed giving a built-up edge indicated by filled-in symbols. The lines in Fig. 11.9 giving combinations of feed and speed satisfying the criteria $T_{\text{mod}} = 700$ K and $T_{\text{int}} = 1000$ K were obtained from the machining theory by the method summarised in Fig. 7.12 using the appropriate work material properties. The temperature factors η and ψ were both taken as 0.7.

For the criteria used the built-up edge range is predicted to fall below the solid lines given in Fig. 11.9 and the experimental results can be seen to give strong support to this prediction. It can be seen that no combinations of feed and speed which gave a built-up edge in the experiments fall above this range while only a few combinations of these factors giving no built-up edge in the experiments fall within it. There is no obvious difference between the experimental results for the tungsten carbide and high speed steel tools at the few points where these can be compared. The predicted and experimental results clearly show that the smaller the feed the higher the speed needed to get rid of built-up edge. This is of course well known in the workshop and has been demonstrated experimentally by many researchers including Heginbotham and Gogia (1961) and Trent (1977). The results in Figs 11.9(a,c) which are both for a rake angle of 5°demonstrate the influence of work material. It can be seen that the predicted lines for $T_{\text{int}} = 1000$ K for the two work materials have different slopes, the 0.36% carbon steel having a higher negative slope than the 0.19% carbon steel. As a result the two lines if plotted together would intersect at a feed of approximately 0.11 mm/rev. Thus when the feed is less than 0.11 mm/rev the line is higher for the 0.36% carbon steel. Hence in this range built-up edge is predicted to persist at higher speeds for the 0.36% carbon steel than for the 0.19% carbon steel. For feeds greater than 0.11 mm this effect is reversed. Unfortunately insufficient experimental results are available in Fig. 11.9(c) to confirm or deny this prediction. The main effect on the predicted results of an increase in rake angle can be seen from Figs. 11.9(b,c) (which are both for the 0.19% carbon steel) to be to increase the negative slope of the $T_{\text{mod}} = 700$ K line and to move it to the right. Thus for certain feeds it is predicted that the cutting speed at which built-up edge disappears increases with increase in rake

---

† It should be noted that in a more complete treatment of the problem the experimentally observed disappearance of built up edge at very low cutting speeds should also be accounted for.

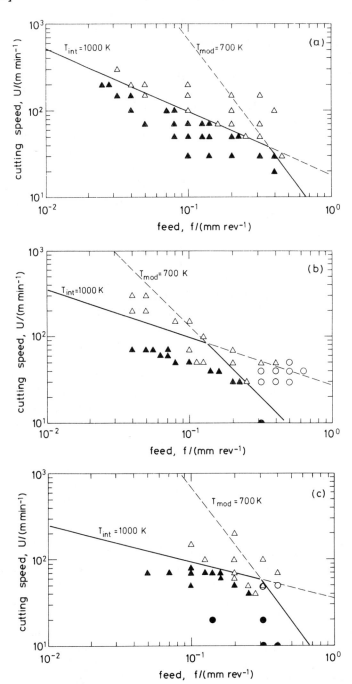

Fig. 11.9 — Predicted and experimental built-up edge conditions: ▲, ●, experimental results for which a built-up edge occurred; △, ○, experimental results for which no built-up edge occurred; ▲, △, experimental results for tungsten carbide tool material; ●, ○, experimental results for high speed steel tool material: (a) $\alpha = 5°$, 0.36% carbon steel; (b) $\alpha = -5°$, 0.19% carbon steel; (c) $\alpha = 5°$, 0.19% carbon steel.

angle and the experimental results in Figs. 11.9(b,c) confirm this. Mathew and Oxley (1980) have also demonstrated this effect of rake angle.

One of the main attributes of free machining steels is that they produce a good surface finish even in the cutting speed range where plain steels give an inferior finish because of built-up edge. To investigate the reasons for this Stevenson and Duncan (1973) made orthogonal turning tests on a resulphurised free machining steel and a plain steel having similar carbon contents. Using temperatures calculated in the way described in section 6.4 from their measured cutting force, tool–chip interface plastic zone thickness and tool–chip contact length values, they determined values of T_{mod} at the interface. Their results showed that when T_{mod} was in the dynamic strain-ageing range a built-up edge was formed for the plain steel but not for the free machining steel. They have given the same explanation as that given in section 7.4 to explain the occurrence of built-up edge with the plain steel. That is, that the stronger material at the interface resulting from dynamic strain ageing forms the built-up edge. They have explained the lack of built-up edge for the free machining steel in the following way. By examining chip sections and the face of the chip which had contacted the tool they found that cracking always occurred along this face, with the cracks evidently emanating from the sulphide inclusions, when T_{mod} was in the dynamic strain ageing range. They reasoned that as a result of this cracking the average strength of the material at the interface would no longer be greater than that of material removed from the interface and therefore that no built-up edge would be formed. It is of interest to note that their analysis of the flow stress in the chip formation zone showed no significant differences in flow stress for the two materials. This therefore seriously questions the suggestion by workers such as Usui and Shaw (1962) and Tipnis and Cook (1965) that the lowering of the average shear strength of the material in the chip formation zone as a result of inclusions is the mechanism which accounts for the free machining properties of resulphurised steels.

Although the dynamic strain ageing effect has been used to help explain the occurrence of built-up edge when machining plain carbon steels it clearly cannot be used to explain built-up edge for materials which do not exhibit this property. It is evident that a built-up edge will tend to occur whenever the distribution of flow stress in the plastic zone at the tool–chip interface is such that the interface is no longer the plane of minimum strength. Since flow stress will in general vary with strain, strain-rate and temperature it can be postulated that for some conditions these parameters will combine in such a way as to produce weaker material away from the interface and thus to cause a built-up edge to be formed. In an investigation of the machining of commercially pure aluminium Bao and Stevenson (1975) determined the strain, strain-rate and temperature distributions and hence the flow stress distribution in the interface plastic zone from experimental flow fields using a method similar to that of Tay *et al.* (1974) when machining without a built-up edge and found that the flow stress over the entire zone was very nearly constant. From this they concluded that a uniform distribution of flow stress in the interface plastic zone is characteristic of steady-state machining without a built-up edge. By extending their analysis to the built-up edge range for the material considered they found that the flow stress in the interface plastic zone tended to decrease in moving away from the cutting face into the chip. They concluded that for these conditions shearing would take place within the chip rather than at the interface and associated this with built-up edge formation.

Such effects will clearly have to be taken into account in extending the machining theory to predict built-up edge conditions for a wider range of materials than those so far considered.

It was mentioned at the start of this section that in order to obtain a good surface finish it is necessary not only to avoid built-up edge but also to ensure that cracks do not occur near to the cutting edge and extend into the newly machined surface. Field and Merchant (1949) using the shear plane model of chip formation (see Chapter 2) reasoned that the chip would be formed by plastic deformation so long as the shear strain occurring at the shear plane was less than the fracture strain. Following Bridgman (1943) they assumed that the amount of shear strain a material can withstand before fracture is linearly dependent on the value of the hydrostatic stress acting in the deforming region. From this they suggested the following fracture criterion for chip formation

$$\frac{\cos \alpha}{\sin \phi \cos (\phi - \alpha)} \geq \gamma_f + s p_a \tag{11.6}$$

where the term on the left-hand side of the equation is the shear strain which occurs as material crosses the shear plane given by equation (2.4), γ_f is the fracture shear strain when the hydrostatic stress is zero, s is the slope of the linear shear fracture strength–hydrostatic stress relation, and p_a is the average value of hydrostatic stress (mean normal compressive stress) acting on the shear plane. Field and Merchant concluded that if equation (11.6) was satisfied before steady-state conditions were reached then a discontinuous chip would result; alternatively, if the steady-state value of ϕ gave a strain less than the fracture strain then the chip would be continuous. Experimental values of shear strain and average hydrostatic stress at the transition from continuous to discontinuous chip formation given by Field and Merchant show only limited agreement with equation (11.6). It should be noted that according to equation (11.6) the chip is either continuous or discontinuous and the well-known experimental observation of cracks only occurring in the chip near to the cutting edge cannot be accounted for. Enahoro and Oxley (1961) have considered this problem. They made orthogonal planing tests at very low cutting speeds in which the only variable was the undeformed chip thickness. Their results showed that for small values of t_1 a continuous chip was formed but that as t_1 was increased cracks appeared in the chip near to the cutting edge, eventually spreading right through the chip as t_1 was further increased. They analysed the stresses in the chip formation zone in a similar way to Palmer and Oxley (see section 3.4) and found that the cracking which occurred with increase in t_1 could be associated with the tensile value of the hydrostatic stress near to the cutting edge which also increased with increase in t_1. They concluded that equation (11.6) could well be a good predictor of the onset of cracking if p_a in this equation was replaced by p_B, the value of the hydrostatic stress at the cutting edge. Foot (1965) has investigated this possibility by making orthogonal planing and turning tests in which he identified the conditions where cracking first occurred for a wide range of cutting conditions. He used the parallel-sided shear zone machining theory of Oxley and Welsh (1963) described in section 4.2 to determine p_B and the average strain-rate in the chip formation zone from his experimental shear angle, etc., results. His results show that when p_B is used in place of p_a then equation (11.6) is a reasonable predictor of the onset of cracking for a given cutting speed. He found, however, that as speed was increased the strain that could be sustained without fracture occurring increased and associated this with the increase in temperature with increase in speed. It is clear from Foot's results that equation (11.6) will need modification to allow for the influence of temperature, etc., before it can be applied with confidence to predict the onset of cracking.

11.6 FURTHER MACHINABILITY CONSIDERATIONS

Although from the viewpoint of surface finish it is desirable to machine with a continuous chip it is also desirable that the chip is subsequently broken into reasonably short lengths for ease of removal. To this end many chip-breaking devices have been developed. The main function of these appears to be to provide an obstruction adjacent to the cutting edge which will bend the chip so that it comes into contact at its free end with either the unmachined work surface or the tool clearance face. The resulting bending moment then breaks the chip. Jawahir (1986) in a comprehensive study of chip breaking has concluded that the most widely used of modern chip breakers, namely the groove chip breaker in which there is a groove in the cutting face set just back from the cutting edge, provides the necessary obstruction by causing the chip to flow back into the groove by the restricted contact mechanism shown in Fig. 2.7. To ensure that the chip will flow back in this way the length of cutting face between the cutting edge and the start of the groove must be less than the natural contact length between chip and tool, i.e. the contact length which would apply when cutting with a plane cutting face tool. The machining theory could therefore be useful in determining the natural contact length for different cutting conditions and thus help in deciding on the positioning of the groove. Much more help could be provided if the machining theory could be extended to the chip formation model for restricted contact conditions given in Fig. 2.7. If this could be achieved then the actual angle of flow into the groove could be predicted and the optimum size and shape of the groove theoretically determined. Information could also be gained on tool stresses and temperatures and estimates therefore made of tool wear and of tool deformation conditions. Many groove chip breaker tools have lumps on the cutting face which apparently act as a second line of defence obstructions. The correct positioning of these lumps requires a knowledge of the chip flow direction across the cutting face and the work described in Chapters 9 and 10 should be helpful in this respect.

It has long been known in the workshop that when machined, under the same cutting conditions, a cold drawn bar of low carbon steel will have better machinability characteristics than a bar of nominally the same chemical composition which has been hot rolled. In the case of the cold drawn material the chip is removed by plastic deformation and an excellent surface finish is produced. With the hot rolled material fracture occurs near the cutting edge as the chip is formed and this results in a rough machined surface. Laboratory investigations of the effects of cold working have shown not only that surface finish can be improved by cold working but also that cutting forces can be reduced and tool life increased. Yaguchi (1985) has considered the influence of cold working on the machinability of a leaded low carbon free machining steel. His experimental results show that the cutting force in the direction of cutting is significantly lower for cold drawn than for hot rolled material with the difference varying from about 15% at a cutting speed of 30 m/min to 5% at 200 m/min. This somewhat paradoxical result of reducing cutting forces by cold working a material so that its hardness is increased has been explained by Kopalinsky and Oxley (1987) using the machining theory suitably modified to take account of cold working effects. They have done this by rewriting equation (4.14) in the form

$$\sigma = \sigma_1 (\varepsilon_w + \varepsilon)^n \qquad (11.7)$$

where ε_w is the effective strain resulting from cold working and ε is the additional strain occurring above this value. It is still assumed that the equations in Appendix A3 represent the flow stress properties σ_1 and n but now in calculating σ prior cold working is taken into account by using equation (11.7) instead of equation (4.14). This approach neglects the differences in the deformation conditions in the cold working stage and subsequent machining operation. Nevertheless, it has the merit of changing the work material properties with cold working in the right direction; that is, an increase in ε_w will increase the strength of the material and effectively reduce its strain-hardening rate. The use of equation (11.7) in place of equation (4.14) leads to the following changes in the machining theory. Equation (7.6) becomes

$$\gamma_{AB} = \gamma_w + \gamma'_{AB} \qquad (11.8)$$

where γ_w is equal to $\sqrt{3}\varepsilon_w$ and γ'_{AB} is given by equation (7.6), and equations (7.4), (7.13) and (7.24) become respectively

$$\tan\theta = 1 + 2\left(\frac{\pi}{4} - \phi\right) - \frac{Cn}{1 + \gamma_w/\gamma'_{AB}} \qquad (11.9)$$

$$h = \frac{t_1 \sin\theta}{\cos\lambda \sin\phi}\left\{1 + \frac{Cn/(1+\gamma_w/\gamma'_{AB})}{3[1 + 2(\pi/4 - \phi) - Cn(1 + \gamma_w/\gamma'_{AB})]}\right\} \qquad (11.10)$$

and

$$\frac{\sigma'_N}{k_{AB}} = 1 + \frac{\pi}{2} - 2\alpha - \frac{2Cn}{1 + \gamma_w/\gamma'_{AB}} \qquad (11.11)$$

The method of obtaining predicted results is the same as that given in the chart in Fig. 7.12 with, however, equations (11.8), (11.9), (11.10) and (11.11) replacing equations (7.4), (7.6), (7.13) and (7.24). In their calculations Kopalinsky and Oxley took the temperature factors η and ψ as equal to 0.7. They made calculations for a range of steels (0.05–0.5% C) and cutting conditions (α = -5° to 5°; U = 25 to 420 m/min; t_1 = 0.125 to 0.5 mm) and allowed the degree of cold working to vary in the range

$\varepsilon_W = 0$–3.0. Their results show that for steels of carbon content less than 0.4% cold working decreases the cutting force in the direction of cutting F_C for most of the conditions considered. They explained this result by referring to the curves of τ_{int} and k_{chip} which gave their equilibrium solutions ($\tau_{int} = k_{chip}$) for ϕ — see Fig. 7.18 for an example of such curves. They observed that the effect of cold working is to raise the τ_{int} curve relative to the k_{chip} curve and hence to move the intersection point of the curves to the right, thus giving a higher value of ϕ and consequently a thinner chip. They found that although cold working not surprisingly increases k_{AB} the accompanying increase in ϕ is usually sufficient to reduce F_C for steels of carbon content less than 0.4%. For steels having a higher carbon content ϕ is still predicted to increase with cold working but this increase is no longer sufficient to compensate for the accompanying increase in k_{AB} and as a consequence F_C increases. To date the modified machining theory has not been applied to predict the influence of cold working on built-up edge formation, tool life, etc.

So far, as mentioned in Chapter 1, the machining theory has not been applied to predict the influence of cutting fluids on machining processes. It is now generally accepted that the main role of a cutting fluid in processes such as turning is to act as a coolant with the main objectives being the cooling of the tool and workpiece, the latter to minimise thermal expansion and distortion which can lead to loss of dimensional accuracy of the machined component. It has been seen above that in turning where tool temperatures are high the main wear mechanism appears to be diffusion. If a cutting fluid can therefore lower the tool temperature the wear rate of the tool should be lowered and the effective tool life increased. Lowering of the tool temperature should also help inhibit deformation of the cutting edge. In slow speed processes such as broaching and tapping it is observed that the cutting fluid can reach the tool–chip interface and act as a lubricant, thus lowering the frictional force and as a consequence the cutting forces. Not surprisingly the cutting fluids selected to act as coolants (water-based fluids) are different from those selected to act as lubricants (neat oils). Armarego and Brown (1969), Trent (1977) and Shaw (1984) have given excellent accounts based mainly on experimental observations of the role of cutting fluids in machining. It is clear from these that the influence of cutting fluids on the chip formation process is very complex and that it would be of considerable help if the machining theory could be extended to take account of the effects of cutting fluids. For the moment it must suffice to consider how in principle this might be achieved. If in processes such as tapping etc. the oil can act at the tool–chip interface so that boundary lubrication conditions exist then presumably the frictional conditions can be described by a mean angle of friction λ as in the shear plane solutions (see Chapter 2) and in the early forms of the present machining theory (see Chapter 4). In this case it should be possible to measure λ either from a machining test or from an independent test and thus to make predictions of cutting force, etc., with relative ease. Even if boundary lubricated conditions are not established over the full contact length between chip and tool it is known from experiments that they usually apply over the final (elastic) part of the contact. It is only by considering this region, which has been ignored in developing the present machining theory, that effects such as the very thick chips obtained by Rowe and Smart (1963) when machining *in vacuo* can be explained. To cover this more complex situation it will be necessary to extend the machining theory to take account not only of the plastic region at the interface but also of the elastic region. In this way the angle of friction describing the frictional conditions in the elastic region will become one of the given conditions and will influence the predicted values of chip thickness, etc. Apparently the influence of this region on the chip formation process is usually small and only becomes important for extreme conditions such as those used by Rowe and Smart. The cooling action of cutting fluids would appear to be most readily allowed for in the machining theory by varying the temperature factors η and ψ in equations (7.7) and (7.10). The way in which cutting temperatures and hence η and ψ vary in the presence of coolants could be investigated as suggested by Tay (1973) using the finite element method referred to in section 6.2, provided that the heat transfer coefficients over the areas affected by the coolant can be estimated. (Muraka *et al.* (1979) have made a preliminary analysis of this kind.) Alternatively, experimental measurements of temperature could be made when machining with coolants and empirical relations obtained for η and ψ. In considering the importance of allowing in the machining theory for the influence of the cooling action of cutting fluids it is worth pointing out that it is generally accepted that above cutting speeds of about 120 m/min cutting fluids have little effect on temperatures and hence on the chip formation process.

When machining 'difficult to machine' materials such as nickel-based alloys it is found that pre-heating of the region of the workpiece which is to be cut as it approaches the tool, e.g. by using a plasma torch, can greatly improve machinability. In this connection it is interesting to note that the machining theory, by including the initial temperature of the work as given information, offers the possibility of investigating the hot machining characteristics of work materials.

11.7 CONCLUDING REMARKS

The machinability predictions described in this chapter have been limited to orthogonal or approximately orthogonal cutting conditions. It should be relatively simple to extend this work to oblique cutting conditions and to bar turning operations using tools with and without a nose radius based on the analyses given in Chapters 8, 9 and 10. Some headway has also been made in analysing the mechanics of more complicated machining processes and this should eventually enable machinability predictions to be made for these processes. In this connection Kopalinsky (1982) has used a statistical description of the rake angle and undeformed chip thickness distributions involved in grinding and has used an early version of the machining theory to predict the cutting forces and temperatures associated with the individual grit cutting edges. By summing these in a suitable way estimates have been made of grinding forces and it has been shown how wheel wear influences these. Using the machining theory in more or less the form represented by the chart in Fig. 7.12 Watson (1985a,b,c,d) and Young (1986) have developed methods for predicting the forces involved in drilling (using a twist drill) and in a simple face milling operation respectively. The basis of both these analyses was to treat the total cut as a series of small elements for which the cutting forces could be found from the machining theory with these elemental forces then summed to give the total forces. An important feature of Watson's analysis is the allowance made for the dynamic cutting geometry of the drill, it of course being no longer acceptable to ignore the feed velocity in defining the cutting velocity, particularly in the region of the chisel edge. Watson has shown reasonable agreement between his predicted and experimental results. In face milling the cutting geometry is changing throughout the cut and in order to apply the machining theory which is for steady-state conditions it must be assumed that the cutting forces reach their steady-state values, appropriate to a given cutting geometry, instantaneously. For the conditions considered in his experiments Young showed experimentally that this assumption was closely met. Excellent agreement was shown by Young between his predicted and experimental forces both in magnitude and in the shape of the force traces (force pulses) during a cut.

To date attention has mainly been limited to plain carbon steel work materials. To extend the work to other materials it will be necessary to obtain the high strain-rate, high temperature flow stress properties for these materials needed in applying the machining theory. At present there is a dearth of such information. An effective way of obtaining the required properties would be by using machining tests as described in Chapter 6. Alternatively, high speed compression tests such as those of Oyane *et al.* (1967), the results of which have been widely applied in this book, or high speed torsion tests such as those described by Stevenson (1975) could be used. For machining applications high speed compression tests would seem to hold an advantage over high speed torsion tests. This is because the plastic deformation in a compression test takes place in the presence of a compressive hydrostatic stress which is also the case for most, if not all, of the plastic deformation occurring in the continuous chip formation process. Thus the well-known tendency for a hydrostatic compressive stress to inhibit the formation of microcracks during plastic deformation applies to both situations. On the other hand in a torsion test, other than when an axial compressive load is applied, microcracks are more easily formed and as a consequence the measured flow stress is reduced below the value it would have in the absence of microcracks. Having argued the superiority of the compression test in this regard it should be pointed out that some workers including Shaw (1984) consider that microcracks can play an important part in determining the flow stress in machining. It is clear that far more research will be needed to determine the most appropriate method for measuring the flow stress properties of work materials and also to replace the relatively crude methods used in this book for relating such properties to the machining process. For example, to what extent is the neglect of history effects justified?

Finally, it might be asked how the chip formation model used as a basis for the machining theory developed in this book can be improved. Perhaps the most obvious shortcoming is the lack of attention given to chip curl. If this could be incorporated into the analysis it would provide further relationships between the deformation in the chip formation zone and along the tool–chip interface in much the way that Dewhurst's rigid–perfectly plastic model of chip formation in Fig. 2.6 does. It might then be expected that the distributions of stress at the tool–chip interface could be obtained as part of the solution and not assumed on the basis of experimental work as at present. Little account has been taken in the machining theory of tool material properties. In future work allowance should be made for the influence of the tool material thermal properties and the shape of the tool as a heat sink when determining cutting temperatures, possibly using numerical methods such as those of Tay *et al.* (1974, 1976) referred to in section 6.2. Account should also be taken of the physical condition of the tool cutting face, e.g. its surface finish which might be expected to vary the sliding velocity at the interface, by the mechanism discussed in section 7.3, and thus the interface temperature and as a consequence chip geometry, cutting forces, etc. In the machining theory it has been assumed that the tool is perfectly sharp, which of course is never the case. To account for the forces resulting from the lack of sharpness it will be necessary to extend the chip formation model to take account of the flow around the cutting edge in a way similar to that shown by the experimental slipline fields in Figs. 3.6 and 3.7. Challen and Oxley (1984b) have already shown how this can be achieved in the case of a rigid–perfectly plastic work material. Once the flow around the cutting edge can be accurately modelled then it will be possible to predict the influence of cutting conditions on the plastic strain put into the newly machined surface and the resulting residual stresses, an understanding of which is becoming increasingly important. As mentioned in section 1.4 some materials such as titanium and stainless steel machine with a chip formation process radically different from the continuous chip formation process assumed in the present theory. In order to develop a predictive machining theory for these materials it will first be necessary to provide an acceptable chip formation model. In this connection it is interesting to report that Manyindo and Oxley (1986) have recently presented a model for the catastrophic shear type of chip obtained when machining stainless steel. This successfully accounts for the saw tooth outer profile of the chip and the adjacent regions of heavily and lightly strained material within the chip and should form a good starting point for the development of a predictive theory.

Appendix A1
Elements of plane strain slipline field theory

A1.1 PLANE STRAIN EQUATIONS REFERRED TO CARTESIAN COORDINATES

Stress definitions and sign convention

Let σ_x, σ_y and σ_z be the direct stresses in the x, y and z directions. Let $\tau_{xy} = -\tau_{yx}$, $\tau_{yz} = -\tau_{zy}$ and $\tau_{zx} = -\tau_{xz}$ be the shear stresses with, for example, τ_{xy} the shear stress acting in the x direction on planes normal to the y axis. Direct stresses are taken positive when tensile, and shear stresses are taken positive when they exert a clockwise couple on the element on which they act.

Definition of plane strain

A state of plane strain is assumed to exist when the flow is parallel to a given plane. Let this be the xy plane with no flow in the z direction. Then if v_x, v_y and v_z are the velocities in the x, y and z directions, $v_z = 0$ and v_x and v_y are independent of z. The basic equations for plane strain conditions defined in this way are now considered.

Strain-rates

The strain-rate components are given by

$$\dot{\varepsilon}_x = \frac{\partial v_x}{\partial x} \qquad \dot{\gamma}_{xy} = \frac{\partial v_y}{\partial x} + \frac{\partial v_x}{\partial y}$$

$$\dot{\varepsilon}_y = \frac{\partial v_y}{\partial y} \qquad \dot{\gamma}_{yz} = \frac{\partial v_z}{\partial y} + \frac{\partial v_y}{\partial z} = 0 \qquad \begin{cases} v_y \neq f(z) \\ v_z = 0 \end{cases} \qquad (A1.1)$$

Sec. A1.1] Plane strain equations referred to Cartesian coordinates

$$\dot{\varepsilon}_z = \frac{\partial v_z}{\partial z} = 0 \quad \dot{\gamma}_{zx} = \frac{\partial v_x}{\partial z} + \frac{\partial v_z}{\partial x} = 0 \quad \begin{cases} v_x \neq f(z) \\ v_z = 0 \end{cases}$$

where $\dot{\varepsilon}_x$, $\dot{\varepsilon}_y$ and $\dot{\varepsilon}_z$ are the direct strain-rates and $\dot{\gamma}_{xy}$, $\dot{\gamma}_{yz}$ and $\dot{\gamma}_{zx}$ are the shear strain-rates.

Stress–strain relations
The work material is assumed to be an isotropic, rigid–plastic material which deforms in accordance with the Lévy–Mises relations. These can be written in the form

$$\frac{\dot{\varepsilon}_x}{\sigma'_x} = \frac{\dot{\varepsilon}_y}{\sigma'_y} = \frac{\dot{\varepsilon}_z}{\sigma'_z} = \frac{\dot{\gamma}_{xy}/2}{\tau_{xy}} = \frac{\dot{\gamma}_{yz}/2}{\tau_{yz}} = \frac{\dot{\gamma}_{zx}/2}{\tau_{zx}} \tag{A1.2}$$

where σ'_x, σ'_y and σ'_z are the deviatoric stress components in the x, y and z directions given by

$$\begin{aligned} \sigma'_x &= \sigma_x - \sigma_m \\ \sigma'_y &= \sigma_y - \sigma_m \\ \sigma'_z &= \sigma_z - \sigma_m \end{aligned} \tag{A1.3}$$

where $\sigma_m = \tfrac{1}{3}(\sigma_x + \sigma_y + \sigma_z)$ is the mean or hydrostatic component of stress. It follows from equations (A1.1) and (A1.2) that $\sigma'_z = \tau_{yz} = \tau_{xz} = 0$. Therefore the z direction is a principal direction with, as shown by the third equation of equations (A1.3),

$$\sigma_z = \tfrac{1}{2}(\sigma_x + \sigma_y) \tag{A1.4}$$

Yield criterion
It follows from equation (A1.4) that both the maximum shear stress and shear strain energy yield criteria can be written as

$$\tfrac{1}{4}(\sigma_x - \sigma_y)^2 + \tau_{xy}^2 = k^2 \tag{A1.5}$$

where k is the shear flow stress. If the maximum shear stress criterion is assumed then $k = \sigma/2$, while if the shear strain energy criterion is assumed $k = \sigma/\sqrt{3}$, where σ is the corresponding uniaxial flow stress.

Equilibrium equations
Neglecting body forces the equilibrium equations are

$$\frac{\partial \sigma_x}{\partial x} + \frac{\partial \tau_{xy}}{\partial y} = 0$$

$$\frac{\partial \sigma_y}{\partial y} + \frac{\partial \tau_{xy}}{\partial x} = 0$$

(A1.6)

Constant volume condition
For the assumed rigid–plastic material elastic strains are taken as zero and during deformation the volume of an element does not change. This constant volume condition can be expressed as

$$\dot{\varepsilon}_x + \dot{\varepsilon}_y + \dot{\varepsilon}_z = 0$$

which for plane strain ($\dot{\varepsilon}_z = 0$) reduces to

$$\dot{\varepsilon}_x + \dot{\varepsilon}_y = 0$$

or, from equations (A1.1),

$$\frac{\partial v_x}{\partial x} + \frac{\partial v_y}{\partial y} = 0 \qquad (A1.7)$$

Isotropic condition
It is implicit in equations (A1.2) that the principal directions of stress and strain-rate or the directions of maximum shear stress and maximum shear strain-rate coincide. From the Mohr stress and strain-rate circles for plane strain conditions, which are given in Fig. A1.1, this is equivalent to $\psi = \psi'$. In terms of stress and strain-rate components

$$\tan(2\psi) = \frac{\sigma_y - \sigma_x}{2\tau_{xy}} \qquad (A1.8)$$

and

$$\tan(2\psi') = \frac{\dot{\varepsilon}_y - \dot{\varepsilon}_x}{\dot{\gamma}_{xy}} \qquad (A1.9)$$

Therefore, equating equations (A1.8) and (A1.9),

$$\frac{\sigma_y - \sigma_x}{2\tau_{xy}} = \frac{\dot{\varepsilon}_y - \dot{\varepsilon}_x}{\dot{\gamma}_{xy}} \qquad (A1.10)$$

or from equations (A1.1)

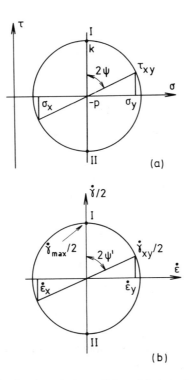

Fig. A1.1 — Mohr's circles for plane strain conditions: (a) stress circle; (b) strain-rate circle.

$$\frac{\sigma_y - \sigma_x}{2\tau_{xy}} = \frac{\partial v_y/\partial y - \partial v_x/\partial x}{\partial v_y/\partial x + \partial v_x/\partial y} \quad \text{(A1.11)}$$

The five equations (A1.5), (A1.6), A1.7) and (A1.11) in terms of the five unknowns σ_x, σ_y, τ_{xy}, v_x and v_y form the basis for solving plane strain plasticity problems.

A1.2 PLANE STRAIN EQUATIONS REFERRED TO SLIPLINES
Definition of sliplines
The sliplines consist of two orthogonal families of curves whose directions at every point in the plastic region coincide with the directions of maximum shear stress and maximum shear strain-rate. To distinguish the sliplines they will be called I and II sliplines with the I lines those on which the shear stress is positive, i.e. exerts a clockwise couple on the element on which it acts. Consider now a small curvilinear element bounded by two pairs of neighbouring sliplines as shown in Fig. A1.2. The state of stress on this small element is represented by the Mohr circle in Fig. A1.1(a). The shear stress on the sliplines is k and the normal stress on the sliplines is

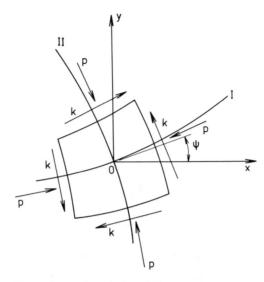

Fig. A1.2 — Curvilinear element bounded by sliplines showing stresses acting on sliplines.

the mean compressive (hydrostatic) stress p where $p = -\frac{1}{2}(\sigma_x + \sigma_y)$. It is evident that the state of stress at a point is completely specified if the values of p and k and the orientation of the sliplines are known.

Stress transformations
From the Mohr stress circle (Fig. A1.1(a)) the transforming equations from the Cartesian coordinate system to the curvilinear slipline system are given by

$$\begin{aligned}\sigma_x &= -p - k\sin(2\psi) \\ \sigma_y &= -p + k\sin(2\psi) \\ \tau_{xy} &= k\cos(2\psi)\end{aligned} \quad \text{(A1.12)}$$

where ψ is the angular rotation of the I lines from the x axis measured positive as shown in Fig. A1.2. Substituting from equations (A1.12) in the equilibrium equations (A1.6) and differentiating and collecting terms gives

$$-\frac{\partial p}{\partial x} - \left(\frac{\partial k}{\partial x} + 2k\frac{\partial \psi}{\partial y}\right)\sin(2\psi) - \left(2k\frac{\partial \psi}{\partial x} - \frac{\partial k}{\partial y}\right)\cos(2\psi) = 0$$

$$-\frac{\partial p}{\partial y} + \left(\frac{\partial k}{\partial y} - 2k\frac{\partial \psi}{\partial x}\right)\sin(2\psi) + \left(2k\frac{\partial \psi}{\partial y} + \frac{\partial k}{\partial x}\right)\cos(2\psi) = 0$$

Sec. A1.2] Plane strain equations referred to sliplines

Selecting the I and II slipline directions to coincide with the x and y directions, i.e. putting $\psi = 0$, yields

$$\frac{\partial p}{\partial s_1} + 2k\frac{\partial \psi}{\partial s_1} - \frac{\partial k}{\partial s_2} = 0 \quad \text{along a I line}$$

$$\frac{\partial p}{\partial s_2} - 2k\frac{\partial \psi}{\partial s_2} - \frac{\partial k}{\partial s_1} = 0 \quad \text{along a II line}$$

(A1.13)

where s_1 and s_2 are distances measured along the I and II lines respectively. The application of these equations which allow for hardening effects is considered in Chapter 3.

In classical slipline field theory the work material is assumed to be perfectly plastic (non-hardening) with the shear flow stress k constant. For k constant, equations (A1.13) reduce to

$$\frac{\partial p}{\partial s_1} + 2k\frac{\partial \psi}{\partial s_1} = 0 \quad \text{along a I line}$$

$$\frac{\partial p}{\partial s_2} - 2k\frac{\partial \psi}{\partial s_2} = 0 \quad \text{along a II line}$$

and these equations can be integrated to give the well-known Hencky equations, i.e.

$$p + 2k\psi = \text{constant along a I line}$$

$$p - 2k\psi = \text{constant along a II line}$$

(A1.14)

These are simply the equilibrium equations referred to sliplines. In general the values of the constants will vary from slipline to slipline.

Velocity transformations

To transform the velocity equations in a similar way let u and v be the velocities along the I and II lines. The transformation equations can then be written

$$v_x = u \cos \psi - v \sin \psi$$

$$v_y = u \sin \psi + v \cos \psi$$

(A1.15)

Substituting from equations (A1.15) in the expressions $\dot{\varepsilon}_x = \partial v_x/\partial x$ and $\dot{\varepsilon}_y = \partial v_y/\partial y$ and differentiating and collecting terms gives

$$\dot{\varepsilon}_x = -\left(u\frac{\partial \psi}{\partial x} + \frac{\partial v}{\partial x}\right)\sin\psi + \left(\frac{\partial u}{\partial x} - v\frac{\partial \psi}{\partial x}\right)\cos\psi$$

$$\dot{\varepsilon}_y = \left(\frac{\partial u}{\partial y} - v\frac{\partial \psi}{\partial y}\right)\sin\psi + \left(\frac{\partial v}{\partial y} + u\frac{\partial \psi}{\partial y}\right)\cos\psi$$

Again selecting the I and II slipline directions to coincide with the x and y directions, i.e. putting $\psi = 0$, and noting that for volume constancy the direct strain-rate along sliplines must be zero (Fig. A1.1(b)) yields

$$\begin{aligned} du - v\,d\psi &= 0 \quad \text{along a I line} \\ dv + u\,d\psi &= 0 \quad \text{along a II line} \end{aligned} \tag{A1.16}$$

These are the Geiringer equations which simply express the property that for volume constancy the rate of extension along sliplines is zero.

A1.3 OUTLINE OF METHOD FOR OBTAINING PERFECTLY PLASTIC SOLUTIONS

General

In steady-state processes such as the machining process considered in the present book continuity of flow considerations impose restrictions on the shape and position of the boundary of the deforming region. The type of problem to be solved is therefore statically undetermined and the solutions for stresses and velocities have to be carried out together. This can be contrasted with a statically determined problem where there are sufficient stress boundary conditions to determine the plastic region and associated stresses with the velocities calculated independently at the end. Slipline field solutions for steady motion problems are therefore obtained by what is essentially a process of trial and error. It is usual to need as a starting point at least some indication of the location and extent of the deforming region. Physical intuition is therefore involved and a number of researchers have developed the construction of slipline fields to what amounts to an art form. In recent years the obtaining of solutions has been facilitated by the development of computer-aided methods but even with these some initial guesswork is usually involved.

Having estimated the approximate position of the boundaries which separate plastically deforming from rigid work material, which must be made up of sliplines (Hill 1950), it is usual to start construction of the slipline field within the deforming region from a known stress boundary condition such as stress-free surface or a tool–work interface along which the frictional condition is known. The construction in this way of a field which is satisfactory for stress can be greatly aided by using slipline field configurations which are known to be internally stress consistent. Such configurations are considered below. Once a field is obtained which is satisfactory for stress it must be checked for velocity by determining the velocities throughout the field

Sec. A1.3] Outline of method for obtaining perfectly plastic solutions

starting at a suitable boundary. If it is not acceptable in this sense then the original field must be modified until both stress and velocity conditions are satisfied. To be finally acceptable it must be shown that the rate of plastic work is positive throughout the field and that a stress distribution exists in the assumed rigid material such that the yield stress is nowhere exceeded.

Slipline field geometry

Hencky has derived a number of geometrical properties of slipline fields which are extremely useful in building-up solutions. For example, one such property can be deduced from the application of equations (A1.14) to a curvilinear element bounded by sliplines as shown in Fig. A1.3. Assuming the hydrostatic stress p'_A is known at

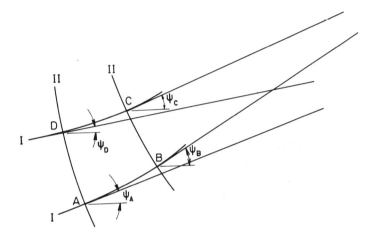

Fig. A1.3 — Curvilinear element bounded by sliplines used in deriving Hencky's first theorem.

point A and the values of ψ are known at points A, B, C and D, then the value of the hydrostatic stress at point C, p'_C, can be calculated in one of two ways using equations (A1.14). This leads to a useful relationship between the inclinations ψ at the points A, B, C and D as follows.

Applying equations (A1.14) from A to B (I line) and then from B to C (II line) gives for p'_C

$$p'_C = p'_A + 2k(\psi_A - 2\psi_B + \psi_C)$$

Alternatively, p'_C can be calculated by applying equations (A1.14) from A to D (II line) and then from D to C (I line), giving for the hydrostatic stress p'_C at C

$$p'_C = p'_A + 2k(-\psi_A + 2\psi_D - \psi_C)$$

As the stress at point C must be single valued, then both expressions for p'_C must be identical and so

$$\psi_A - 2\psi_B + \psi_C = -\psi_A + 2\psi_D - \psi_C$$

and therefore

$$\psi_C - \psi_B = \psi_D - \psi_A \qquad (A1.17)$$

This equation which is known as Hencky's first theorem shows that there must be a constant angle between the tangents to two sliplines of one family at their intersections with sliplines of the other family. Therefore, any two orthogonal families of curves which satisfy this property are possible elements for use in constructing slipline fields. Two obvious examples are (i) two orthogonal families of straight lines and (ii) a centred fan made up of straight lines and circular arcs. These have formed the basis of many slipline field solutions.

Stress boundary conditions
Two stress boundary conditions which are frequently encountered in constructing slipline fields are stress-free surfaces and tool surfaces along which the frictional shear stress opposing motion of the work material is known. The Mohr stress circle in Fig. A1.4 shows that the sliplines must meet a free surface at an angle of $\pi/4$ with

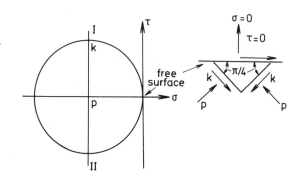

Fig. A1.4 — Mohr's stress circle and slipline field element showing stress-free surface condition.

$p = k$. In the example given p is compressive. If p is tensile then the signs of the shear stresses are also reversed. If the frictional shear stress is τ then it is easily shown that the sliplines must meet the tool–work interface at an angle θ given by $\theta = \frac{1}{2}\cos^{-1}(\tau/k)$. From this it can be seen that if $\tau = 0$ the sliplines meet the interface at an angle of $\pi/4$ while if $\tau = k$ the sliplines are normal and tangential to the interface.

Sec. A1.3] **Outline of method for obtaining perfectly plastic solutions** 217

Hodograph
To check a slipline field for velocity it is quite common to construct a velocity diagram, termed a hodograph, instead of using equations (A1.16) for this purpose. A hodograph is constructed on the basis that in order to make the rate of extension along sliplines zero (volume constancy condition) adjacent points on a slipline must have a relative velocity which cuts the line joining them at right-angles. Two examples are given in Fig. A1.5. In the first (Fig. A1.5(a)) the velocity v_p is assumed

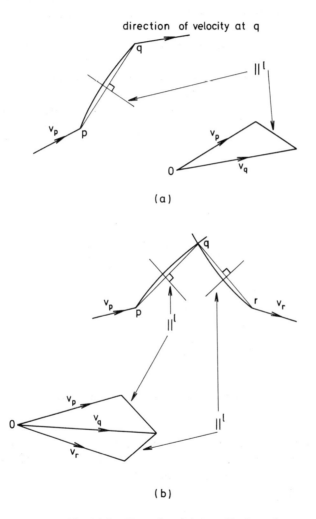

Fig. A1.5 — Examples of elemental hodographs.

known at a point p on the slipline pq while at q the direction but not the magnitude of the velocity is known. The magnitude is determined by selecting its value to make the velocity of p relative to q normal to pq as shown. In constructing this elemental hodograph velocities have been measured from O (often referred to as the pole of the hodograph), and in the example given pq has been approximated by an equivalent chord. In the second example (Fig. A1.5(b)) sliplines pq and qr intersect at point q with the velocities known at p and r but not at q. The velocity of q is found by selecting v_q to make the relative velocities normal to the corresponding sliplines as shown. It can be seen that if small enough steps are considered then corresponding elements of the slipline field and hodograph will be orthogonal. As volume constancy also applies in the case of rigid–plastic hardening materials the above examples are also applicable to the construction of hodographs for hardening slipline fields such as those given in Chapter 3.

In rigid–perfectly plastic slipline field analyses for metal working processes it is usual to represent what in real materials would be a zone of intense shear by a slipline across which there is a discontinuity in the tangential component of velocity. This is acceptable as the only constraint resulting from the volume constancy condition is that the normal component of velocity should be continuous across the slipline. Such a case is shown in Fig. A1.6 where the velocity is changed instantaneously from U to

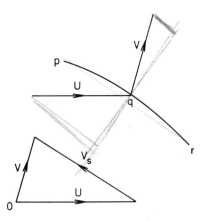

Fig. A1.6 — Elemental hodograph for tangential velocity discontinuity.

V in crossing the slipline pr at point q which requires a jump (discontinuity) in the tangential component of velocity equal to V_S. It follows from the appropriate equation of equations (A1.16) that V_S is constant along pr.

Further reading
For a more rigorous treatment of slipline field theory than that given here readers are referred to Hill's (1950) masterly account of the theory. In this he starts by showing that the differential system governing the plane strain deformation of a rigid–perfectly plastic material is hyperbolic with the sliplines being the characteristics of the system. This approach has a number of advantages. It provides, for example, an elegant proof that the boundary between the plastic deforming region and the rigid

Sec. A1.3] Outline of method for obtaining perfectly plastic solutions

region consists of sliplines. The actual equations for the variation of stress and velocity along sliplines obtained by Hill are of course no different from those obtained here. Excellent descriptions of slipline field theory and its application have also been given by Ford and Alexander (1977), Johnson and Mellor (1983), Johnson *et al.* (1982) and Kachanov (1974).

Appendix A2
Relating plane strain and uniaxial conditions and determining the shear flow stress distribution for experimental flow fields

In the slipline field analyses of slow cutting speed experimental flow fields described in Chapter 3 the velocity distribution was known for a given field and it was required to find the associated distribution of shear flow stress k. In general, it is obvious that the strain distribution must first be estimated and then from some form of the stress–strain curve for the work material the corresponding values of k found. Such a process presupposes that it is possible to use the stress–strain curve obtained from a simple test on the material to express the relative stress and strain for a complex system of stress. To achieve this it is usual to introduce the concept of an effective stress and an effective natural strain.

The effective stress for a three-dimensional stress state may be defined as

$$\bar{\sigma} = \frac{\sqrt{2}}{2}[(\sigma_1-\sigma_2)^2+(\sigma_2-\sigma_3)^2+(\sigma_3-\sigma_1)^2]^{1/2} \qquad (A2.1)$$

where σ_1, σ_2 and σ_3 are the principal stresses. The corresponding invariant of strain-rate, the effective strain-rate, may be defined as

$$\dot{\bar{\varepsilon}} = \frac{\sqrt{2}}{3}[(\dot{\varepsilon}_1-\dot{\varepsilon}_2)^2+(\dot{\varepsilon}_2-\dot{\varepsilon}_3)^2+(\dot{\varepsilon}_3-\dot{\varepsilon}_1)^2]^{1/2} \qquad (A2.2)$$

where $\dot{\varepsilon}_1$, $\dot{\varepsilon}_2$ and $\dot{\varepsilon}_3$ are the principal strain-rates. This expression may be integrated to give the effective natural strain, i.e.

$$\bar{\varepsilon} = \int \dot{\bar{\varepsilon}} \, dt = \frac{\sqrt{2}}{3} \int [(\dot{\varepsilon}_1 - \dot{\varepsilon}_2)^2 + (\dot{\varepsilon}_2 - \dot{\varepsilon}_3)^2 + (\dot{\varepsilon}_3 - \dot{\varepsilon}_1)^2]^{1/2} dt \qquad (A2.3)$$

It is then assumed that for a given material the effective stress is a unique function of the effective natural strain.

In uniaxial compression (the test used to obtain the stress–strain curves in the slipline field analyses considered) the effective stress reduces to the single principal stress and, as straining is proportional and the principal axes remain unchanged, the effective strain-rate can be integrated and reduces to the principal natural strain. Thus a stress–strain curve obtained from a compression test can be regarded as an effective stress–effective natural strain curve. This assumes, of course, that in the compression test the strain is measured as $ln(h_0/h)$ where h_0 and h are the original and deformed heights respectively of the specimen.

In plane strain the effective stress reduces to

$$\bar{\sigma} = \sqrt{3} \tau_{max} \qquad (A2.4)$$

where τ_{max} is the maximum shear stress, which during flow is equivalent to the shear flow stress k, and the effective natural strain to

$$\bar{\varepsilon} = \frac{1}{\sqrt{3}} \int \dot{\gamma}_{max} \, dt \qquad (A2.5)$$

where $\dot{\gamma}_{max}$ is the maximum shear strain-rate. It was not possible to perform this integration to give an expression in terms of the total natural direct and shear strains in the slipline field analyses of the experimental flow fields because the principal directions changed during straining. However, as can be seen from the Mohr strain-rate circle in Fig. A1.1(b), $\dot{\gamma}_{max}$ can be expressed as

$$\dot{\gamma}_{max} = (\dot{\gamma}_{xy}^2 + 4\dot{\varepsilon}_x^2)^{1/2}$$

or substituting for $\dot{\gamma}_{xy}$ and $\dot{\varepsilon}_x$ from equations (A1.1)

$$\dot{\gamma}_{max} = \left[\left(\frac{\partial v_y}{\partial x} + \frac{\partial v_x}{\partial y} \right)^2 + 4 \left(\frac{\partial v_x}{\partial x} \right)^2 \right]^{1/2} \qquad (A2.6)$$

and as the velocities v_x and v_y and hence the velocity gradients $\partial v_x/\partial x$ etc. could be measured from the experimental flow fields it was possible to determine $\dot{\gamma}_{max}$ and hence, by graphical integration along streamlines, $\bar{\varepsilon}$. Once $\bar{\varepsilon}$ was known at a point in the plastic zone then the corresponding value of $\bar{\sigma}$ could be determined and hence k found from equation (A2.4). It should be noted that the directions of maximum shear strain-rate at a point in the plastic zone can also be determined from the measured velocity gradients from equation (A1.9).

In the slipline field analyses described in Chapter 3 flow stress was assumed to vary only as a result of strain-hardening. However, in the predictive machining theory and in comparing flow stress results obtained from machining tests with those obtained from high speed compression and tension tests account was also taken of the influence of strain-rate and temperature on flow stress. Plane strain and uniaxial conditions were still related using an effective stress and strain but it was now assumed that for a given material the effective stress would only be a unique function of the effective strain for given values of effective strain-rate and temperature or, combining these parameters, of velocity-modified temperature. If σ, ε and $\dot{\varepsilon}$ are taken as the uniaxial stress, strain and strain-rate values and k, γ and $\dot{\gamma}$ as the plane strain maximum shear stress, shear strain and shear strain-rate values then following the above definitions of effective stress etc. these can be related as follows:

$$k = \frac{\sigma}{\sqrt{3}}$$
$$\gamma = \sqrt{3}\varepsilon \quad \text{(A2.7)}$$
$$\dot{\gamma} = \sqrt{3}\dot{\varepsilon}$$

Appendix A3
Representation of σ_1 and n as functions of velocity-modified temperature and carbon content for a range of plain carbon steels

The original reference of Oyane *et al.* (1967) gives full stress–strain data for the 0.16% carbon steel but only a summary of results for the other materials tested, namely 0.33%, 0.49% and 0.52% carbon steels. Professor M. Oyane kindly provided full details for these three materials to Dr. W. F. Hastings in a private communication in 1973. The following treatment of the data follows that given by Hastings (1975).

Values of σ_1 and n for the 0.33%, 0.49% and 0.52% carbon steels determined in the same way as described in section 6.5 are given in Figs. A3.1 and A3.2 together

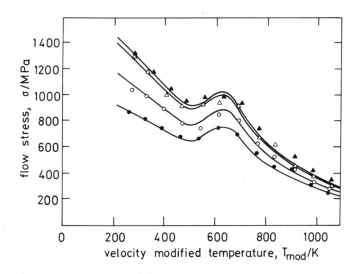

Fig. A3.1 — Values of σ_1 obtained from high speed compression test results: ▲, 0.52% carbon steel; △, 0.49% carbon steel; ○, 0.33% carbon steel; ●, 0.16% carbon steel.

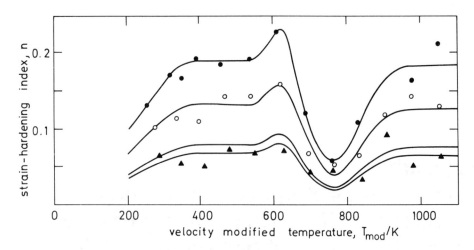

Fig. A3.2 — Values of n obtained from high speed compression test results: symbols same as in Fig. A3.1.

with the results obtained for the 0.16% carbon steel. The n results for the 0.49% carbon steel which mainly lie between those for the 0.33% and 0.52% carbon steels have been omitted for clarity. The σ_1 and n results (Figs. A3.1 and A3.2) can be seen to follow the same general trends with a clear dynamic strain ageing range for all of the steels considered. From the σ_1 results (Fig. A3.1) it is evident that the principal effect of increasing the carbon content is to raise the flow stress over the entire range of T_{mod} with the increase becoming less with increase in T_{mod}. The n results (Fig. A3.2) on the other hand show a decrease in n with increase in carbon content. Noting the regular behaviour of the σ_1 and n results with, for a given T_{mod}, both σ_1 and n changing more or less linearly with change in carbon content, Hastings attempted a general description of the results for all of the steels considered. He did this by using curves fitted to the σ_1 and n results for the 0.16% carbon steel as a basis together with rescaling functions to allow for the influence of carbon content.

The curves for σ_1 and n for the 0.16% carbon steel were described mathematically in the following way. The values of σ_1 and n were obtained at T_{mod} intervals of 25 K from smooth curves drawn through the data points. These values were used in a polynomial regression program which first plotted the input data and calculated the required regression curve and then plotted this curve over the original data plot. Printed output from the program included the regression coefficients. The order of the polynomial and the range of T_{mod} were adjusted to obtain close agreement between the data from the original curves and those computed from the regression program. To obtain the required accuracy sections of seventh-order polynomials had to be used except for certain parts of the curves where the data were best fitted by straight lines. Because of the iterative technique to be employed in the machining analysis it was found necessary to match all sections of the curves as closely as possible. This was achieved by obtaining the slope of adjacent sections and determin-

Appendix A3

ing the values of T_{mod} at which the minimum difference in slope occurred, thus fixing the value of T_{mod} at which the change from one equation to the next occurred. The values of σ_1 and n at the change-over points were determined for each curve and the constant in the equation for the right-hand section was adjusted to make the two values agree. The correction to the constant was in all cases small, being of the order of 1% of the calculated values.

The equations for σ_1 (MPa) and n obtained in this way can be written as follows.

For σ_1:

$$\sigma_1 = 1126.62 - 0.98421 T_{mod}$$
for $T_{mod} \leq 458$

$$\sigma_1 = -19914.15 + 135.07 T_{mod} - 0.20137 T_{mod}^2$$
$$-3.1090 \times 10^{-4} T_{mod}^3 + 7.2551 \times 10^{-7} T_{mod}^4$$
$$+7.3255 \times 10^{-10} T_{mod}^5 - 2.2977 \times 10^{-12} T_{mod}^6$$
$$+1.2673 \times 10^{-15} T_{mod}^7$$
for $458 < T_{mod} \leq 748$

$$\sigma_1 = 17756.97 - 97.198 T_{mod} + 0.23022 T_{mod}^2$$
$$-2.4637 \times 10^{-4} T_{mod}^3 + 2.8921 \times 10^{-8} T_{mod}^4$$
$$+1.8495 \times 10^{-10} T_{mod}^5 - 1.6072 \times 10^{-13} T_{mod}^6$$
$$+4.2722 \times 10^{-17} T_{mod}^7$$
for $748 < T_{mod} \leq 1200$

$$\sigma_1 = 172.42$$
for $T_{mod} > 1200$

(A3.1)

For n:

$$n = 0.04768$$
for $T_{mod} \leq 73$

$$n = 0.04937 + 3.5861 \times 10^{-4} T_{mod} - 1.4026 \times 10^{-5} T_{mod}^2$$
$$+1.7680 \times 10^{-7} T_{mod}^3 - 9.4992 \times 10^{-10} T_{mod}^4$$
$$+2.7341 \times 10^{-12} T_{mod}^5 - 4.1361 \times 10^{-15} T_{mod}^6$$
$$+2.5569 \times 10^{-18} T_{mod}^7$$
for $73 < T_{mod} \leq 396$

$$n = 0.19109$$
for $396 < T_{mod} \leq 528$

$$n = -145.26 + 0.81927 T_{mod}$$
$$-0.88538 \times 10^{-3} T_{mod}^2 - 2.5350 \times 10^{-6} T_{mod}^3$$
$$+5.0364 \times 10^{-9} T_{mod}^4 + 2.4501 \times 10^{-12} T_{mod}^5$$
$$-1.04279 \times 10^{-14} T_{mod}^6$$

(A3.2)

$$+5.8410\times10^{-18}T_{mod}^{7}$$
for $528 < T_{mod} \leqslant 693$

$$n = -21.227 + 0.08507 T_{mod}$$
$$-4.4837\times10^{-5}T_{mod}^{2} - 1.3310\times10^{-7}T_{mod}^{3}$$
$$-3.5910\times10^{-11}T_{mod}^{4} + 5.1253\times10^{-13}T_{mod}^{5}$$
$$-5.1724\times10^{-16}T_{mod}^{6} + 1.5471\times10^{-19}T_{mod}^{7}$$
for $693 < T_{mod} \leqslant 827$

$$n = -65.632 + 0.30193 T_{mod}$$
$$-0.49548\times10^{-3}T_{mod}^{2} + 2.7300\times10^{-7}T_{mod}^{3}$$
$$+9.1267\times10^{-11}T_{mod}^{4} - 1.0362\times10^{-13}T_{mod}^{5}$$
$$-3.1959\times10^{-17}T_{mod}^{6} + 3.0674\times10^{-20}T_{mod}^{7}$$
for $827 < T_{mod} \leqslant 974$

$$n = 0.18388$$
for $T_{mod} > 974$

The rescaling functions for σ_1 and n were initially obtained from a combination of the results given in Figs. A3.1 and A3.2 and some σ_1 and n results obtained from stress–strain curves given by Zorev (1966) for a 0.11% carbon steel and a 0.03% carbon Armco iron. Zorev's results were obtained from slow speed compression tests at room temperature using incremental loading techniques. As no information regarding strain-rate or temperature was given for these tests it was assumed that reasonable values would be 0.01/s and 20°C respectively. Neglecting any temperature rise in the tests this gave a T_{mod} value for the compression tests of approximately 350 K. Using values of σ_1 and n obtained from Figs. A3.1 and A3.2 for this value of T_{mod} together with Zorev's results it was shown that the equations

$$\sigma_1 = 531.31 + 753.17C \tag{A3.3}$$

$$n = 0.244 - 0.3396C \tag{A3.4}$$

where C is the percentage carbon content and the units of σ_1 are MPa, represented the variations of σ_1 and n with carbon content accurately. If now these equations are divided by the value of σ_1 and n for the 0.16% carbon steel at $T_{mod} = 350$ K then the rescaling functions

and
$$\text{SIGFAC} = (531.31 + 753.17C)/651.72 \tag{A3.5}$$

$$\text{NFAC} = (0.244 - 0.3396C)/0.189 \tag{A3.6}$$

are obtained. It was at first hoped that the values of σ_1 and n could be obtained for any of the steels in the range considered by simply multiplying the values for the 0.16% carbon steel at the given value of T_{mod} by these rescaling functions. Although this was found to be the case for n it was found that two additional functions dependent on both carbon content and T_{mod} were necessary to obtain an adequate description of the σ_1 results. These were selected by trial and error to give as good a fit as possible with the experimental results in Fig. A3.2. The functions obtained in this way can be written in the form

$$\sigma_1 = \sigma_{1b}[1+(\text{SIGFAC}-1)(1400-T_{\text{mod}})/900]$$
$$\text{for } 200 \leqslant T_{\text{mod}} \leqslant 700$$
$$\sigma_1 = \sigma_{1b}[1+C(T_{\text{mod}}-700)/T_{\text{mod}}]$$
$$\text{for } 700 < T_{\text{mod}} \leqslant 1100$$

(A3.7)

where σ_{1b} is the value of σ_1 for the 0.16% carbon steel at a given value of T_{mod}.

For a given carbon content and T_{mod}, σ_1 and n can now be determined. For the given T_{mod}, σ_1 is found for the 0.16% carbon steel from equations (A3.1) and substituted as σ_{1b} in equations (A3.7) to obtain the value of σ_1 corresponding to the same T_{mod} and the given carbon content. In the case of n the appropriate value for the 0.16% carbon steel is found from equations (A3.2) and then multiplied by the value of NFAC found from equation (A3.6) using the given carbon content. The lines in Figs. A3.1 and A3.2 represent results found in this way and can be seen to give an encouragingly good fit with the original experimental results.

References

Ansell, C. T., and Taylor, J. (1962) The surface finishing properties of a carbide and ceramic cutting tool. In: *Proceedings 3rd International Machine Tool Design and Research Conference*. Pergamon, Oxford, pp. 225–243.

Armarego, E. J. A., and Brown, R. H. (1962) On the size effect in metal cutting. *Int. J. Prod. Res.* **1**, 75–99.

Armarego, E. J. A., and Brown, R. H. (1969) *The Machining of Metals*. Prentice-Hall, Englewood Cliffs, NJ.

Armarego, E. J. A., and Wiriyacosol, S. (1978a) Oblique machining with triangular form tools—I. Theoretical investigation. *Int. J. Mach. Tool Des. Res.* **18**, 67–80.

Armarego, E. J. A., and Wiriyacosol, S. (1978b) Oblique machining with triangular form tools—II. Experimental investigation. *Int. J. Mach. Tool Des. Res.* **18**, 153–165.

Backer, W. R., Marshall, E. R., and Shaw, M. C. (1952) The size effect in metal cutting. *Trans. ASME* **74**, 61–71.

Bagchi, A., and Wright, P. K. (1987) Stress analysis in machining with the use of sapphire tools. *Proc. R. Soc. Lond.* **A409**, 99–113.

Bao, H., and Stevenson, M. G. (1975) An investigation of built-up edge formation in the machining of aluminium. *Int. J. Mach. Tool. Des. Res.* **16**, 165–178.

Bao, H., Montgomery, J. R., and Stevenson, M. G. (1976) Performance measurements and improvements of a quick-stop device for metal cutting studies. University of New South Wales, Industrial Engineering Department Report.

Bever, M. B., Marshall, E. R., and Tichenor, L. B. (1953) The energy stored in metal chip during orthogonal cutting. *J. Appl. Phys.* **24**, 1117–1119.

Blok, H. (1938) Theoretical study of temperature rise at surfaces of actual contact under oiliness lubricating conditions. In: *Proceedings of the General Discussion on Lubrication and Lubricants*. Institution of Mechanical Engineers, pp. 222–235.

Boothroyd, G. (1963) Temperatures in orthogonal metal cutting. *Proc. Inst. Mech. Eng.* **177**, 789–802.

Boothroyd, G. (1965) *The Fundamentals of Metal Machining*. Edward Arnold,

References

London.

Boothroyd, G., Eagle, J. M., and Chisholm, A. W. J. (1967) Effect of tool flank wear on the temperatures generated during metal cutting. In: *Proceedings 8th International Machine Tool Design and Research Conference*. Pergamon, Oxford, pp. 667–680.

Bowden, F. P., and Tabor, D. (1950) *The Friction and Lubrication of Solids*. Clarendon, Oxford.

Bridgman, P. W. (1943) On torsion combined with compression. *J. Appl. Phys.* **14**, 273–283.

Brown, R. H., and Armarego, E. J. A. (1964) Oblique machining with a single cutting edge. *Int. J. Mach. Tool Des. Res.* **4**, 9–25.

Bruce, P. W. (1972) Tool wear evaluation using theoretically predicted tool tip temperature. B.E. Thesis, University of New South Wales.

Campbell, J. D., and Ferguson, W. G. (1970) The temperature and strain-rate dependence of the shear strength of mild steel. *Phil. Mag.* **21**, (8th Series), 63–81.

Challen, J. M., and Oxley, P. L. B. (1979) An explanation of the different regimes of friction and wear using asperity deformation models. *Wear* **53**, 229–243.

Challen, J. M., and Oxley, P. L. B. (1984a) A slip line field analysis of the transition from local asperity contact to full contact in metallic sliding friction. *Wear* **100**, 171–193.

Challen, J. M., and Oxley, P. L. B. (1984b) Slip-line fields for explaining the mechanics of polishing and related processes. *Int. J. Mech. Sci.* **26**, 403–418.

Challen, J. M., McLean, L. J., and Oxley, P. L. B. (1984) Plastic deformation of a metal surface in sliding contact with a hard wedge: its relation to friction and wear. *Proc. R. Soc. Lond.* **A394**, 161–181.

Chandrasekaran, H., and Kapoor, D. V. (1965) Photoelastic analysis of tool–chip interface stresses. *J. Eng. Ind.* **87**, 495–502.

Chao, B. T., and Trigger, K. J. (1955) Temperature distribution at the tool–chip interface in metal cutting. *Trans. ASME* **77**, 1074–1121.

Childs, T. H. C. (1971) A new visio-plasticity technique and a study of curly chip formation. *Int. J. Mech. Sci.* **13**, 373–387.

Childs, T. H. C., and Maekawa, K. (1987) A computer simulation approach towards the determination of optimum cutting conditions. In: *Proc. ASM Int. Conf. on Strategies for Automation of Machining, Orlando*. pp. 157–166.

Chisholm, A. J. (1946) In: Discussion on Machinability. *Proc. Inst. Mech. Eng.* **155**, 267–268 and 278–279.

Chisholm, A. J. (1949) The characteristics of machined surfaces. *Machinery (London)* **74**, 729–736.

CIRP (1976) *International Directory of Machining and Production Information Centres*.

Colding, B. N. (1959) A wear relationship for turning, milling and grinding. *Machining Economics*. Lund Press, Stockholm.

Collins, I. F. (1968) The algebraic geometry of slip-line fields with applications to boundary value problems. *Proc. R. Soc. Lond.* **A303**, 317–338.

Collins, I. F. (1972) A simplified analysis of the rolling of a cylinder on a rigid/perfectly plastic half-space. *Int. J. Mech. Sci.* **14**, 1–14.

Collins, I. F. (1979) The application of singular perturbation theory to the analysis of forming processes for strain-hardening materials. In: *IUTAM Symposium on Metal Forming Plasticity, Tutzing*. Springer, Berlin, pp. 227–243.

Colwell, L. V. (1954) Predicting the angle of chip flow for single point cutting tools. *Trans. ASME* **76**, 199–204.

Conning, S. W., and Oxley, P. L. B. (1988) Visioplasticity. In: Blazynski, T. A. (ed.) *Plasticity and modern metal-forming technology*. Elsevier, Amsterdam.

Conning, S. W., Farmer, L. E., and Oxley, P. L. B. (1984) Investigation of experimental flow patterns for plane strain extrusion of hardening materials with slipline field methods. *Phil. Trans. R. Soc. Lond.* **A311**, 495–522.

Cook, N. H., and Goldberger, D. S. (1982) Machining parameter selection using on-line adaptive control. In: *On the Art of Cutting Metals—75 Years Later*. PED **7**, ASME, pp. 159–166.

Crookall, J. R., and Richardson, D. B. (1969) Use of photographed orthogonal grids and mechanical quick-stopping techniques in machining research. In: *Photography in Engineering*. Institution of Mechanical Engineers, London, pp. 27–36.

Dewhurst, P. (1978) On the non-uniqueness of the machining process. *Proc. R. Soc. Lond.* **A360**, 587–610.

Dewhurst, P. (1979) The effect of chip breaker constraints on the mechanics of the machining process. *CIRP Annals* **28**, 1–5.

Dewhurst, P., and Collins, I. F. (1973) A matrix technique for constructing slip-line field solutions to a class of plane-strain plasticity problems. *Int. J. Num. Meth. Eng.* **7**, 357–378.

Doyle, E. D., Horne, J. G., and Tabor, D. (1979) Frictional interactions between chip and rake face in continuous chip formation. *Proc. R. Soc. Lond.* **A366**, 173–183.

Dutt, R. P., and Brewer, R. C. (1964) On the theoretical determination of the temperature field in orthogonal machining. *Int. J., Prod. Res.* **4**, 91–114.

Eggleston, D. M., Herzog, R., and Thomsen, E. G. (1959) Observations on the angle relationships in metal cutting. *J. Eng. Ind.* **81**, 263–279.

Enahoro, H. E., and Oxley, P. L. B. (1961) An investigation of the transition from a continuous to a discontinuous chip in orthogonal machining. *Int. J. Mech. Sci.* **3**, 145–156.

Enahoro, H. E., and Oxley, P. L. B. (1966) Flow along tool–chip interface in orthogonal metal cutting. *J. Mech. Eng. Sci.* **8**, 36–41.

Ernst, H. (1938) Physics of metal cutting. *Machining of Metals*. American Society for Metals.

Ernst, H., and Merchant, M. E. (1941) Chip formation, friction and high quality machined surfaces. *Trans. Am. Soc. Met.* **29**, 299–378.

Farmer, L. E., and Fowle, R. F. (1979) An experimental procedure for studying the flow in plane strain extrusion. *Int. J. Mech. Sci.* **21**, 599–608.

Fenton, R. G., and Oxley, P. L. B. (1967) Predicting cutting forces at super high cutting speeds from work material properties and cutting conditions. In: *Proceedings 8th International Machine Tool Design and Research Conference*. Pergamon, Oxford, pp. 247–258.

Fenton, R. G., and Oxley, P. L. B. (1968–1969) Mechanics of orthogonal machining: allowing for the effects of strain-rate and temperature on tool–chip friction.

Proc. Inst. Mech. Eng. **183**, 417–438.

Fenton, R. G., and Oxley, P. L. B. (1969–1970) Mechanics of orthogonal machining: predicting chip geometry and cutting forces from work-material properties and cutting conditions. *Proc. Inst. Mech. Eng.* **184**, 927–942.

Field, M., and Merchant, M. E. (1949) Mechanics of formation of the discontinuous chip in metal cutting. *Trans. ASME* **71**, 421–430.

Finnie, I. (1956) Review of the metal cutting analyses of the past hundred years. *Mech. Eng.* **78**, 715–721.

Foot, R. M. (1965) A study of conditions at the transition boundary between continuous and discontinuous chip formation. Diploma in Advanced Engineering, Cranfield College of Aeronautics.

Ford, H., and Alexander, J. M. (1977) *Advanced Mechanics of Materials*. Ellis Horwood, Chichester, 2nd edn.

Goriani, V. L., and Kobayashi, S. (1967) Strain and strain-rate distributions in orthogonal metal cutting. *CIRP Annals* **15**, 425–431.

Green, A. P. (1955) Friction between unlubricated metals: a theoretical analysis of the junction model. *Proc. R. Soc. Lond.* **A228**, 191–204.

Hahn, R. S. (1951) On the temperature developed at the shear plane in metal cutting. In: *Proc. 1st U.S. Nat. Cong. of Appl. Mech.* pp. 661–666.

Hastings, W. F. (1967) A new quick-stop device and grid technique for metal cutting research. *CIRP Annals* **15**, 109–116.

Hastings, W. F. (1975) A theoretical and experimental investigation of the machining process. Ph.D. Thesis, University of New South Wales.

Hastings, W. F., and Oxley, P. L. B. (1976) Predicting tool life from fundamental work material properties and cutting conditions. *CIRP Annals* **25**, 33–38.

Hastings, W. F., Mathew, P., and Oxley, P. L. B. (1980) A machining theory for predicting chip geometry, cutting forces, etc., from work material properties and cutting conditions. *Proc. R. Soc. Lond.* **A371**, 569–587.

Hastings, W. F., Stevenson, M. G., and Oxley, P. L. B. (1974) Predicting a material's machining characteristics using flow stress properties obtained from high-speed compression tests. *Proc. Inst. Mech. Eng.* **188**, 245–252.

Hastings, W. F., Mathew, P., Oxley, P. L. B., and Taylor, J. (1979) Estimated cutting temperatures—their use as a predictor of tool performance when machining plain carbon steels. In: *Proceedings 20th International Machine Tool Design and Research Conference*. Macmillan, London, pp. 313–320.

Heginbotham, W. B., and Gogia, S. L. (1961) Metal cutting and the built-up nose. *Proc. Inst. Mech. Eng.* **175**, 892–905.

Hill, R. (1950) *The Mathematical Theory of Plasticity*. Clarendon, Oxford.

Hill, R. (1951) On the state of stress in a plastic-rigid body at the yield point. *Phil. Mag.* **42**, (7th Series), 868–875.

Hill, R. (1954) The mechanics of machining: a new approach. *J. Mech. Phys. Solids.* **3**, 47–53.

Hu, R. S., and Mathew, P. (1983) A revised theory for predicting cutting forces in oblique machining. University of New South Wales, Industrial Engineering Department Report.

Hu, R. S., Mathew, P., Oxley, P. L. B., and Young, H. T. (1986) Allowing for end cutting edge effects in predicting forces in bar turning with oblique machining conditions. *Proc. Inst. Mech. Eng., Part C* **200** (C2), 89–99.

Jahja, I. (1986) An experimental and theoretical investigation of the flow at the tool–chip interface in machining. M.Eng.Sc. Thesis, University of New South Wales.

Jawahir, I. S. (1986) An experimental and theoretical study of the effects of tool restricted contact on chip breaking. Ph.D. Thesis, University of New South Wales.

Johnson, R. W., and Rowe, G. W. (1967–1968) Bulge formation in strip drawing with light reduction in area. *Proc. Inst. Mech. Eng.* **182**, 521–529.

Johnson, W. (1962) Some slip-line fields for swaging or expanding, indenting, extruding and machining for tools with curved dies. *Int. J. Mech. Sci.* **4**, 323–347.

Johnson, W. (1967) Cutting with tools having a rounded edge—some theoretical considerations. *CIRP Annals* **14**, 315–319.

Johnson, W., and Mellor, P. B. (1983) *Engineering Plasticity*. Ellis Horwood, Chichester.

Johnson, W., Sowerby, R., and Venter, R. D. (1982) *Plane-strain Slip-line Fields for Metal-deformation Processes: a Source Book and Bibliography*, Pergamon, Oxford.

Kachanov, L. M. (1974) *Fundamentals of the Theory of Plasticity*. MIR, Moscow.

Kato, S., Yamaguchi, K., and Yamada, M. (1972) Stress distribution at the interface between tool and chip in machining. *J. Eng. Ind.* **94**, 683–688.

Kececioglu, D. (1958) Shear strain-rate in metal cutting and its effects on shear-flow stress. *Trans. ASME* **80**, 158–168.

Klopstock, H. (1925) Recent investigations in turning and planing and a new form of cutting tool. *Trans. ASME* **47**, 345–377.

Kluft, W., Konig, W., van Luttervelt, C. A., Nakayama, K., and Pekelharing, A. J. (1979) Present knowledge of chip control. *CIRP Annals* **28** (2), 441–455.

Kobayashi, S., and Thomsen, E. G. (1959) Some observations on the shearing process in metal cutting. *J. Eng. Ind.* **81**, 251–262.

Kobayashi, S., and Thomsen, E. G. (1960) The role of friction in metal cutting. *J. Eng. Ind.* **82**, 324–332.

Kobayashi, S., and Thomsen, E. G. (1962) Metal cutting analysis—II New parameters. *J. Eng. Ind.* **84**, 71–80.

Koenigsberger, F. (1964) *Design Principles of Metal-cutting Machine Tools*. Pergamon, Oxford.

Kopalinsky, E. M. (1982) Modelling of material removal and rubbing processes in grinding as a basis for realistic determination of workpiece temperature distributions. *Wear* **81**, 115–134.

Kopalinsky, E. M., and Oxley, P. L. B. (1984) Size effects in metal removal processes. In: *Inst. Phys. Conf. Ser. 70; 3rd Conf. Mech. Prop. High Rates of Strain*. pp. 389–396.

Kopalinsky, E. M., and Oxley, P. L. B. (1987) Pedicting the effects of cold working on the machining characteristics of low carbon steels. *J. Eng. Ind.* **109**, 257–264.

Kronenberg, M. (1954) *Grundzuge der Zerspanungslehre*. Springer, Berlin, 2nd edn.

Kronenberg, M. (1966) *Machining Science and Application*. Pergamon, Oxford.

Kudo, H. (1965) Some new slip-line solutions for two-dimensional steady-state

machining. *Int. J. Mech. Sci.* **7**, 43–45.
Larson-Basse, J., and Oxley, P. L. B. (1973) Effect of strain-rate sensitivity on scale phenomena in chip formation. In: *Proceedings 13th Machine Tool Design and Research Conference*. Macmillan, London, pp. 209–216.
Lee, E. H., and Shaffer, B. W. (1951) The theory of plasticity applied to a problem of machining. *J. Appl. Mech.* **18**, 405–413.
Leone, W. C. (1954) Distribution of shear zone heat in metal cutting. *Trans. ASME* **76**, 121–125.
Lin, G. C. I. (1978) Prediction of cutting forces and chip geometry in oblique machining from flow stress properties and cutting conditions. *Int. J. Mach. Tool Des. Res.* **18**, 117–130.
Lin, G. C. I., and Oxley, P. L. B., (1972) Mechanics of oblique machining: predicting chip geometry and cutting forces from work material properties and cutting conditions. *Proc. Inst. Mech. Eng.* **186**, 813–820.
Lin, G. C. I., Mathew, P., Oxley, P. L. B., and Watson, A. R. (1982) Predicting cutting forces for oblique machining conditions. *Proc. Inst. Mech. Eng.* **196**, 141–148.
Ling, F. E., and Saibel, E. (1956) On the tool life and temperature relationship in metal cutting. *Trans. ASME* **78**, 1113–1117.
Loladze, T. N. (1962) Adhesion and diffusion wear in metal cutting. *J. Inst. Eng. India, Part ME 2* **43**, 108.
Low, A. H. (1962) Effects of initial conditions in metal cutting. NEL Report 49.
Lowen, E. G., and Shaw, M. C. (1954) On the analysis of cutting tool temperatures. *Trans. ASME* **76**, 217–236.
Luk, W. K. (1969) The direction of chip flow for a single point lathe tool with zero nose radius. *Int. J. Mach. Tool Des. Res.* **9**, 391–399.
Luk, W. K. (1972) The direction of chip flow in oblique cutting. *Int. J. Prod. Res.* **10**, 67–76.
MacGregor, C. W., and Fisher, J. C. (1946) A velocity-modified temperature for the plastic flow of metals. *J. Appl. Mech.* **13**, A11–A16.
Manjoine, M. J. (1944) Influence of rate of strain and temperature on yield stresses of mild steel. *J. Appl. Mech.* **11**, A211–A218.
Manyindo, B. M., and Oxley, P. L. B. (1986) Modelling the catastrophic shear type of chip when machining stainless steel. *Proc. Inst. Mech. Eng.* **200**, 349–358.
Mathew, P. and Oxley, P. L. B. (1980) Predicting the cutting conditions at which built-up edge disappears when machining plain carbon steels. *CIRP Annals* **29**, 11–14.
Mathew, P. and Oxley, P. L. B. (1981) Allowing for the influence of strain hardening in determining the frictional conditions at the tool–chip interface in machining. *Wear* **69**, 219–234.
Mathew, P., Hastings, W. F., and Oxley, P. L. B. (1979) Machining—a study in high strain-rate plasticity. In: *Inst. Phys. Conf. Ser. 47; 2nd Conf. Mech. Prop. High Rates of Strain*. pp. 360–371.
Merchant, M. E. (1944) Basic mechanics of metal cutting process. *J. Appl. Mech.* **11**, A168–A175.
Merchant, M. E. (1945) Mechanics of the metal cutting process. II. Plasticity conditions in orthogonal cutting. *J. Appl. Phys.* **16**, 318–324.

Metcut (1980) *Machining Data Handbook*. Metcut Research Associates Inc., Cincinnati, OH, 3rd edn.

Muraka, P. D., Barrow, G., and Hinduja, S. (1979) Influence of the process variables on the temperature distribution in orthogonal machining using the finite element method. *Int. J. Mech. Sci.* **21**, 445–456.

Nachev, V. I., and Oxley, P. L. B. (1985) Predicting cutting conditions giving plastic deformation of the cutting edge in terms of tool and work material properties. In: *Proceedings 25th International Machine Tool Design and Research Conference*. Macmillan, London, pp. 225–230.

Nakayama, K. (1956) Temperature rise of workpiece during metal cutting. *Bull. Fac. Eng. Yokohama Nat. Univ.* **5**, 1–10.

Nakayama, K. (1959) Studies of the mechanism in metal cutting. *Bull. Fac. Eng. Yokohama Nat. Univ.* **8**, 1–26.

OECD–CIRP (1966) *Proceedings of the Seminar on Metal Cutting*.

Ohmori, M., and Yoshinaga, Y. (1966) The effect of deformation rate on the strength and blue-brittleness temperature of a mild steel. *9th Japan Congress on Testing Materials*. pp. 58–63.

Ohmori, M., and Yoshinaga, Y. (1968) Effect of strain-rate on the compressive deformation resistance of steel. *11th Japan Congress on Testing Materials*. pp. 95–99.

Okushima, K., and Minato, K. (1959) On the behaviour of chip in steel cutting. *Bull. Jap. Soc. Mech. Eng.* **2** (5), 58–64.

Ota, T., Shindo, A., and Fukoaka, H. (1958) An investigation of the theories of orthogonal machining. *Trans. Jap. Soc. Mech. Eng.* **24**, 484–493.

Oxley, P. L. B. (1963a) Note: allowing for friction in estimating upper bound loads. *Int. J. Mech. Sci.* **5**, 183–184.

Oxley, P. L. B. (1963b) Rate of strain effect in metal cutting. *J. Eng. Ind.* **85**, 335–337.

Oxley, P. L. B. (1966) Introducing strain-rate dependent work material properties into the analysis of orthogonal cutting. *CIRP Annals* **13**, 127–138.

Oxley, P. L. B., and Hastings, W. F. (1976) Minimum work as a possible criterion for determining the frictional conditions at the tool/chip interface in machining. *Phil. Trans. R. Soc. Lond.* **282**, 565–584.

Oxley, P. L. B., and Hastings, W. F. (1977) Predicting the strain-rate in the zone of intense shear in which the chip is formed in machining from the dynamic flow stress properties of the work material and the cutting conditions. *Proc. R. Soc. Lond.* **A356**, 395–410.

Oxley, P. L. B., and Stevenson, M. G. (1967) Measuring stress/strain properties at very high strain-rates using a machining test. *J. Inst. Met.* **95**, 308–313.

Oxley, P. L. B., and Welsh, M. J. M. (1963) Calculating the shear angle in orthogonal metal cutting from fundamental stress, strain, strain-rate properties of the work material. In: *Proceedings 4th International Machine Tool Design and Research Conference*. Pergamon, Oxford, pp. 73–86.

Oxley, P. L. B., and Welsh, M. J. M. (1967) An explanation of the apparent Bridgman effect in Merchant's orthogonal cutting results. *J. Eng. Ind.* **89**, 549–555.

Oxley, P. L. B., Humphreys, A. G., and Larizadeh, A. (1961) The influence of rate

of strain-hardening in machining. *Proc. Inst. Mech. Eng.* **175**, 881–891.
Oxley, P. L. B., Hastings, W. F., and Stevenson, M. G. (1974) Predicting cutting forces, tool life, etc., using work material flow stress properties obtained from high speed compression tests. In: *Proceedings of International Conference on Production Engineering, Tokyo.* pp. 528–534.
Oyane, M. (1973) Private communication.
Oyane, M., Takashima, F., Osakada, K., and Tanaka, H. (1967) The behaviour of some steels under dynamic compression. *10th Japan Congress on Testing Materials.* pp. 72–76.
Pal, A. K., and Koenigsberger, F. (1968) Some aspects of the oblique cutting process. *Int. J. Mach. Tool Des. Res.* **8**, 45–57.
Palmer, W. B., and Oxley, P. L. B. (1959) Mechanics of metal cutting. *Proc. Inst. Mech. Eng.* **173**, 623–654.
Palmer, W. B., and Yeo, R. C. K. (1963) Metal flow near the tool point during orthogonal cutting with a blunt tool. In: *Proceedings 4th International Machine Tool Design and Research Conference.* Pergamon, Oxford, pp. 61–71.
Rapier, A. C. (1954) A theoretical investigation of the temperature distribution in the metal cutting process. *Brit. J. Appl. Phys.* **5**, 400–405.
Rosenhain, W., and Sturney, A. C. (1925) Report on flow and rupture of metals during cutting. *Proc. Inst. Mech. Eng.* **114**, 141–174.
Roth, R. N. (1969) Analysis of flow round tool cutting edge. Ph.D. Thesis, University of New South Wales.
Roth, R. N., and Oxley, P. L. B. (1972) A slip-line field analysis for orthogonal machining based on experimental flow fields. *J. Mech. Eng. Sci.* **14**, 85–97.
Rowe, G. W., and Smart, E. F. (1963) The importance of oxygen in dry machining of metal on a lathe. *Brit. J. Appl. Phys.* **14**, 924–926.
Rowe, G. W., and Spick, P. T. (1967) A new approach to determination of the shear-plane angle in machining. *J. Eng. Ind.* **89**, 530–538.
Russel, J. K. (1964) Oblique machining with a single cutting edge. M.Eng.Sc. Thesis, University of Melbourne.
Russel, J. K., and Brown, R. H. (1966) The measurement of chip flow direction. *Int. J. Mach. Tool Des. Res.* **6**, 129–138.
Schallamach, A. (1971) How does rubber slide? *Wear* **17**, 301–312.
Shallbroch, H., and Schaumann, H. (1937) Der schnittemperatur biem drehvorgang und ihre anwondurig als zerspanbarkeitskonziffer. *Zeitschrift VDI* **81**, 325.
Shaw, M. C. (1968) Historical aspects concerning removal operations on metals. In: *Metal Transformations.* Gordon and Breach, New York, pp. 211–260.
Shaw, M. C. (1984) *Metal Cutting Principles.* Clarendon, Oxford.
Shaw, M. C., Cook, N. H., and Finnie, I. (1953) Shear angle relationships in metal cutting. *Trans. ASME* **75**, 273–288.
Shaw, M. C., Usui, E., and Smith, P. A. (1961) Free machining steel—III. Cutting forces, surface finish and chip formation. *J. Eng. Ind.* **83**, 181–193.
Spanns, C. (1970) A systematic approach to three dimensional chip curl, chip breaking and chip control. SME paper MR70-241.
Spanns, C., and van Geel, P. F. H. J. (1970) Break mechanism in cutting with a chip breaker. *CIRP Annals* **18**, 87–92.
Stabler, G. V. (1951) The fundamental geometry of cutting tools. *Proc. Inst. Mech.*

Eng. **165**, 14–21.

Stabler, G. V. (1964) The chip flow law and its consequences. In: *Proceedings 5th International Machine Tool Design and Research Conference*. Pergamon, Oxford, pp. 243–251.

Stevenson, M. G. (1970) A theoretical and experimental investigation of the influence of strain-rate in machining. Ph.D. Thesis, University of New South Wales.

Stevenson, M. G. (1975) Torsional Hopkinson-bar tests to measure stress–strain properties relevant to machining and high speed forming. In: *Proceedings of 3rd North American Metalworking Research Conference, Pittsburgh*. Carnegie Press, pp. 291–304.

Stevenson, M. G., and Duncan, K. R. (1973) Effect of manganese sulphide inclusions on the tool/chip interface shear stress in machining low-carbon steel. *J. Iron Steel Inst.* **211**, 710–717.

Stevenson, M. G., and Oxley, P. L. B. (1969–1970) An experimental investigation of the influence of speed and scale on the strain-rate in a zone of intense plastic deformation. *Proc. Inst. Mech. Eng.* **184**, 561–576.

Stevenson, M. G., and Oxley, P. L. B. (1970–1971) An experimental investigation of the influence of strain-rate and temperature on the flow stress properties of a low carbon steel using a machining test. *Proc. Inst. Mech. Eng.* **185**, 741–754.

Stevenson, M. G., and Oxley, P. L. B. (1973) High temperature stress–strain properties of a low-carbon steel from hot machining tests. *Proc. Inst. Mech. Eng.* **187**, 263–272.

Stevenson, M. G., Bao, H., and Montgomery, J. R. (1978) Measurement of strain-rate in the primary deformation zone in the machining of three metals. University of New South Wales, Industrial Engineering Department Report.

Stevenson, M. G., Wright, P. K., and Chow, J. G. (1983) Further development in applying the finite element method to the calculation of temperature distributions in machining and comparison with experiment. *J. Eng. Ind.* **105**, 149–154.

Takeyama, H., and Murata, R. (1963) Basic investigation of tool wear. *J. Eng. Ind.* **85**, 33–38.

Tanaka, K., and Kinoshita, M. (1967) Compressive strength of mild steel at high strain-rate and at high temperature. *Bull. Japan Soc. Mech. Eng.* **10**, 429–436.

Tay, A. O. (1973) A numerical study of the temperature distribution generated during orthogonal machining. Ph.D. Thesis, University of New South Wales.

Tay, A. O., Stevenson, M. G., and de Vahl Davis, G. (1974) Using the finite element method to determine temperature distributions in orthogonal machining. *Proc. Inst. Mech. Eng.* **188**, 627–638.

Tay, A. O., Stevenson, M. G., de Vahl Davis, G., and Oxley, P. L. B. (1976) A numerical method for calculating temperature distributions in machining from force and shear angle measurements. *Int. J. Mach. Tool Des. Res.* **16**, 335–349.

Taylor, F. W. (1907) On the art of cutting metals. *Trans. ASME* **28**, 31–350.

Taylor, G. I., and Quinney, H. (1934) The latent energy remaining in a metal after cold working. *Proc. R. Soc. Lond.* **A143**, 307–326.

Taylor, G. I., and Quinney, H. (1937) The emission of the latent energy due to previous cold working when a metal is heated. *Proc. R. Soc. Lond.* **A163**, 157–181.

Taylor, J. (1955) Milling cutters with high relief angles. *Machinery (London)* **87**, 1063–1070.
Tipnis, V. A., and Cook, N. H. (1965) Influence of MnS-bearing inclusions on flow and fracture in a machining shear zone. In: *Mechanical Working and Steel Processing, Met. Soc. Conf.* **44**, 285–307.
Tobias, S. A. (1965) *Machine Tool Vibration*. Blackie, Glasgow.
Trent, E. M. (1967) Hot compressive strength of cemented carbides. In: *Proceedings 8th International Machine Tool Design and Research Conference*. Pergamon, Oxford, pp. 629–642.
Trent, E. M. (1977) *Metal Cutting*. Butterworths, Guildford.
Trigger, K. J. and Chao, B. T. (1951) An analytical evaluation of metal cutting temperatures. *Trans. ASME* **73**, 57–68.
Trigger, K. J. and Chao, B. T. (1956) The mechanism of crater wear of cemented carbide tools. *Trans. ASME* **78**, 1119–1126.
Usui, E., and Hirota, A. (1978) Analytical prediction of three dimensional cutting process part 2 chip formation and cutting force with conventional single-point tool. *J. Eng. Ind.* **100**, 229–235.
Usui, E., and Hoshi, K. (1963) Slip-line fields in metal cutting involving centred fan fields. In: *International Research in Production Engineering*, ASME, New York, pp. 61–71.
Usui, E., and Makino, R. (1967) An example of stress and strain distributions in slow speed, steady-state machining. *J. Japan Soc. Prec. Eng.* **33**, 237–245.
Usui, E., and Shaw, M. C. (1962) Free machining steel—IV. Tools with reduced contact. *J. Eng. Ind.* **84**, 89–102.
Usui, E., and Shirakashi, T. (1982) Mechanics of machining—from descriptive to predictive theory. In: *On the Art of Cutting Metals—75 Years Later*. PED **7**, ASME, pp. 13–35.
Usui, E., and Takeyama, H. (1960) A photoelastic analysis of machining stresses. *J. Eng. Ind.* **82**, 303–308.
Usui, E., Hirota, A., and Masuko, M. (1978a) Analytical prediction of three dimensional cutting process part 1 basic cutting model and energy approach. *J. Eng. Ind.* **100**, 222–228.
Usui, E., Shirakashi, T., and Kitagawa, T. (1978b) Analytical prediction of three dimensional cutting process part 3 cutting temperature and crater wear of carbide tool. *J. Eng. Ind.* **100**, 236–243.
van Luttervelt, C. A., and Pekelharing, A. J. (1976) Chip formation in machining operations at small diameter. *CIRP Annals* **25** (1) 71–76.
Wallace, P. W., and Boothroyd, G. (1964) Tool forces and tool–chip friction in orthogonal machining. *J. Mech. Eng. Sci.* **6**, 74–87.
Watson, A. R. (1985a) Geometry of drill elements. *Int. J. Mach. Tool Des. Res.* **25**, 209–227.
Watson, A. R. (1985b) Drilling model for cutting lip and chisel edge and comparison of experimental and predicted results. I—Initial cutting lip model. *Int. J. Mach. Tool Des. Res.* **25**, 347–365.
Watson, A. R. (1985c) Drilling model for cutting lip and chisel edge and comparison of experimental and predicted results. II—Revised cutting lip model. *Int. J. Mach. Tool Des. Res.* **25**, 367–376.

Watson, A. R. (1985d) Drilling model for cutting lip and chisel edge and comparison of experimental and predicted results. III—Drilling model for chisel edge. *Int. J. Mach. Tool Des. Res.* **25**, 377–392.

Weiner, J. H. (1955) Shear-plane temperature distribution in orthogonal cutting. *Trans. ASME* **77**, 1331–1341.

Woolman, J., and Mottram, R. A. (1964) *The Mechanical and Physical Properties of the British En Steels*. British Iron and Steel Research Association, Pergamon, Oxford.

Woxen, R. (1937) Tool-life and balances of heat in lathe work. *Ingeniörsvetenskapsakademien*. Handlingar 142, Stockholm.

Wright, P. K., and Trent, E. M. (1973) Metallographic methods of determining temperature gradients in cutting tools. *J. Iron Steel Inst.* **211**, 364–368.

Wright, P. K., Horne, J. G., and Tabor, D. (1979) Boundary conditions at the tool chip interface in machining: comparison between seizure and sliding friction. *Wear* **54**, 371–390.

Yaguchi, H. (1985) Effect of cold working on the machinability of low-carbon leaded free-machining steel. In: *Proceedings of the 13th North American Manufacturing Research Conference*. pp. 245–251.

Young, H. T. (1986) Application of predictive machining theory to more complicated processes including single-point oblique cutting tools with other than single straight cutting edges and milling, and experimental verification. Ph.D. Thesis, University of New South Wales.

Young, H. T., Mathew, P., and Oxley, P. L. B. (1987) Allowing for nose radius effects in predicting the chip flow direction and cutting forces in bar turning. *Proc. Inst. Mech. Eng., Part C* **201** (C3), 213–226.

Zorev, N. N. (1963) Inter-relationship between shear processes occurring along toolface and shear plane in metal cutting. In: *International Research in Production Engineering*. ASME, New York, pp. 42–49.

Zorev, N. N. (1966) *Metal Cutting Mechanics*. Pergamon, Oxford.

Index

built-up edge,
 criteria other than dynamic strain ageing for occurrence of, 130–131, 197–198, 200–210
 dynamic strain ageing criterion for occurrence of, 129–130, 197–198
 in slipline field solution, 27
 influence of sulphide inclusions on formation of, 200
 influence on surface finish of, 197
 nature of, 19–21
 prediction of cutting conditions for occurrence of and comparison with experimental results, 129–131, 197–200

catastrophic shear type chip model, 207
chip breaking,
 basic mechanisms of, 202
 possibility of predicting best cutting edge geometry for, 202
chip cracking,
 criterion for, 201
 influence on surface finish of, 197, 201
 prediction of cutting conditions for, 201–202
chip flow direction (angle),
 comparison of predicted and experimental results,
  cutting on side cutting edge only, 139–140
  cutting on side and end cutting edges, 151, 158
  cutting with nose radius tools, 173–177
 definition of, 138
 experimental measurement of, 148
 prediction methods,
  cutting on side cutting edge only, 137–140
  cutting on side and end cutting edges, 151–157
  cutting with nose radius tools, 160–173
chip formation models used in developing machining theory,
 model based on experimental flow fields, 97–98
 parallel-sided shear zone model, 50–51
 shear plane model, 23–25
chip formation processes,
 basic chip types, 19–21
 experimental investigations of, 19

chip formation zone stress distributions, 43–45, 48–49, 51–52
chip thickness,
 comparison of predicted and experimental results,
  effect of carbon content, 125, 127–128
  effect of cutting speed, 125–127
  effect of rake angle, 125
 experimental measurement of, 29, 77, 122
 prediction calculation procedure, *see* shear angle calculation procedure
 prediction methods, *see* shear angle prediction methods
 use of determining experimental shear angle values, 29, 77
cold working,
 influence on machinability of, 202–203
 predicting effect of on machinability, 203–204
computer-aided methods, construction and analysis of experimental slipline fields, 49
curled chip slipline field solutions,
 perfectly plastic, 32–34
 strain-hardening, 47–49
cut geometry,
 oblique conditions,
  relation of undeformed chip thickness to feed and side cutting edge angle, 142
  relation of width of cut to tube wall thickness (depth of cut) and side cutting edge angle, 142
 orthogonal conditions,
  undeformed chip thickness, 18
  width of cut, 18
cutting edge deformation,
 comparison of predicted and experimental cutting conditions for, 195–197
 importance of in determining rate of flank wear, 192
 prediction of cutting conditions for, 192–193
cutting edge geometry,
 oblique conditions,
  effective rake angle, definition of, 136
  end cutting edge angle, definition of, 155
  inclination angle, definition of, 136, 155

normal rake angle, definition of, 136, 155
side cutting edge angle, definition of, 142, 155
orthogonal conditions,
  clearance angle, definition of, 19
  rake angle, definition of, 19
cutting fluids,
  neglect of effects of, 18n
  possibility of predicting effects of, 204–205
  role in machining of, 204
cuttting force components and relations between them,
  oblique conditions, 141–143
  orthogonal conditions, 24–25
cutting forces,
  comparison of predicted and experimental results, nose radius tools,
    effect of cutting speed, 179
    effect of depth of cut, 179
    effect of nose radius, 179
  comparison of predicted and experimental results, oblique conditions,
    effect of cutting speed, 143–146
    effect of end cutting edge, 151, 158–159
    effect of inclination angle, 143–146
    effect of normal rake angle, 143–146
    effect of side cutting edge angle, 143–146
  comparison of predicted and experimental results, orthogonal conditions,
    effect of carbon content, 125, 127
    effect of cutting speed, 63–64, 123–127
    effect of rake angle, 64, 123–125
    effect of undeformed clip thickness, 64, 124–125, 127
  experimental measurement of, 29, 77, 120, 147
  prediction calculation procedures,
    nose radius tools, 177
    oblique conditions, 143, 148–150, 158
    orthogonal conditions, see shear angle calculation procedure
  prediction methods,
    nose radius tools, 177
    oblique conditions, 140–143, 158
    orthogonal conditions, see shear angle prediction methods
cutting power, prediction of, 183–184
cutting velocity (speed),
  definition of, 18, 120n
  experimental measurement of, 119–120

dynamic strain ageing,
  as built-up edge criterion, 129–130, 197–198
  effect on flow stress of, 93, 96

equivalent cutting edge,
  description of, 151–157
  use of in predicting chip flow angle, 158
  use of in predicting cutting forces, 158, 177
experimental flow fields,
  methods for obtaining of,
    ciné filming method, 38–39
    'quick-stop' method, 39, 66–67
  general observations from,

chip formation zone's substantial width, 40
cracking at chip formation zone free surface, 43, 48
stagnation point on cutting edge, 40, 42
swept back nature of deformation of chip undersurface and machined surface, 40–42

finite element methods,
  applied to chip formation problem, 22
  applied to temperature calculations, 83–84
friction angle at tool-chip interface.
  definition of, 25
  experimental measurement of, 29
  use of as friction parameter in predicting shear angle, 26–32, 61–62
frictional shear stress at tool-chip interface,
  experimental measurement of, 63, 64, 92
  expressed in terms of velocity modified temperature, 64
  use of as friction parameter in predicting shear angle, 32, 34, 62–64, 102–104

hot machining, possibility of investigating the effect of using machining theory, 205

inertial stresses and forces, neglect of, 22n, 63n

machinability factors, definition of, 183
machining experiments, description of,
  for checking predictions of cutting forces, etc. for orthogonal conditions, 119–121
  for checking shear angle equations, 28–29
  for determining cutting conditions giving built-up edge, 198
  for determining cutting conditions giving cutting edge deformation, 193–195
  for measuring chip flow angle for oblique conditions, 147–148
  for measuring cutting forces for oblique conditions, 142, 147–148
  for measuring cutting forces and chip flow angle for nose radius tools, 173–177
  for measuring size effect, 131–132
  for measuring strain-rates and material properties, 56, 66–67, 76–77
  for obtaining experimental flow fields, 38–40, 66–67
  for selecting feed to give maximum metal removal rate, 191–192
machining processes considered in book, 17
machining research,
  historical, 22
  reasons for doing it, 17–18
maximum metal removal rate,
  experimental measurement of feed for, 191–192
  prediction of feed for, 197

oblique machining process,
  definitions of, 136
orthogonal machining process,
  definition of, 18–19

Index

plane strain conditions in, 22

parallel sided shear zone model, 50–51
plane strain conditions,
  definition of, 208
  in orthogonal machining, 22

restricted tool-chip contact solutions, 34

shear angle,
  comparison of predicted and experimental results,
    apparent Bridgman-type effect in Merchant's results, 62
    effect of cutting speed, 62–63
    effect of strain-hardening, 54–55
    effect of undeformed chip thickness, 62, 132–134
    for shear plane solutions, 30–33
  definition of, 25, 51, 69, 98
  experimental measurement of, 29, 77
  prediction calculation procedures,
    using parallel-sided shear zone theory, 61–63
    using theory based on experimental chip formation model, 102–104, 114–118
  prediction methods,
    parallel-sided shear zone theory, 51–53
    shear-plane and related solutions, 26–35
    theory based on experimental chip formation model, 97–102
shear plane model,
  in oblique machining, 140
  in orthogonal machining, 23–25
shear strain in chip formation,
  occurring at shear plane, 25
  occurring in parallel-sided shear zone, 53
  occurring in chip formation zone determined from experimental flow fields, 70, 73
size effect,
  experimental results for, 124–125, 132–134
  explanation of, 127, 134
slipline field analysis of experimental flow fields,
  basis of visioplasticity technique, 38
  check on stress and force balances, 43–45, 48–49
  construction of slipline fields, 42–43, 45–48
  determination of flow stress distribution, 43, 220–221
  determination of hydrostatic stress distribution, 43–44
slipline fields obtained from analyses of experimental flow fields,
  for chip formation zone, 42, 47
  for complete chip formation process, 47
  for flow along tool-chip interface including stagnation point, 46, 47
slipline field solutions,
  limitations of perfectly plastic (shear plane) solutions, 35–36
  nature of hardening solutions, 37–38
  perfectly plastic slipline fields, 26–28, 32–35

strain-hardening slipline fields based on experimental flow fields, 42–49
specific cutting pressure,
  definition of, 132
  experimental results for, 132
  explanation of increase of with decrease in undeformed chip thickness, 134
strain-hardening,
  effect of on shear angle, 54–55
  importance of in determining hydrostatic stress distributions, 43, 43–44
strain-rate distributions in chip formation zone determined from experimental flow fields,
  effect of cutting speed and undeformed chip thickness on, 70–73
  method of determining of, 67–70
  nature of, 70–72
strain-rate in chip formation zone,
  effect of cutting speed and undeformed chip thickness on, 70–73
  experimental measurement of average value of, 55
  experimental measurement of distributions of, 65–73
  determination of as part of solution, 105–109
strain-rate in tool-chip interface plastic zone,
  determination of as part of solution, 113
  determination of from experimentally measured thickness of plastic zone, 91, 101–102, 104
stress distributions,
  as tool-chip interface, 47, 49, 63, 100, 101, 106
  in chip formation zone, 43–45, 48–49, 51–52, 201
surface finish,
  factors determining best attainable, 197
  influence of built-up edge on, 197
  influence of chip cracking on, 197

temperatures in machining.
  calculation methods and procedures,
    based on experimental flow fields using finite element methods, 83–84
    based on finite chip formation and tool-chip interface plastic zones, 80–83
    based on shear plane model, 78–80
    iterative procedures involved in, 87, 91–92, 102–104, 116–117
  comparison of finite element and experimental results, 83, 84–85
  comparison of finite element results and results obtained from finite plastic zones method, 84–86
  effect of cutting speed on, 87
  experimental measurement of, 80–81, 85
tool-chip contact length,
  calculation of, 63, 84, 101, 202, 203
  experimental measurement of, 45, 47, 84, 91
  restriction of, 34
tool-chip interface friction,
  nature of, 62–63, 101–102, 109–113
  parameters, of,

frictional shear stress, 32, 62–64, 102–104
  mean friction angle, 25
tool-chip interface plastic zone thickness,
  comparison of predicted and experimental results,
  effect of carbon content, 128
  effect of cutting speed, 128–129
  effect of rake angle, 128
  experimental measurement of, 122–123
  prediction calculation procedure, 114–118
  prediction method, 113
tool-chip interface stress distributions,
  assumed normal stress distributions, 63, 101, 106
  determined from experimental flow field analyses, 47, 49, 101
  photoelastic measurements of, 47, 100
tool wear and tool life,
  comparison of predicted and experimental results, 187–190
  definitions of, 184
  diffusion as dominant wear mechanism, 185–186, 189
  empirical relations for, 185
  influence of tool temperature on tool life, 185
  prediction of influence of variations in work material hardness on tool life, 187–189
  predictions of tool life, 186–190

undeformed chip thickness, relation to feed and side cutting edge angle, 66, 142
uniqueness of shear plane solutions,
  experimental investigations of, 34
  question of, 31–32, 32–34

velocity modified temperature,
  definition of, 92
  support for use of in representing material flow stress properties for conditions appropriate to machining, 94
  use of in comparing experimental results, 92–93
velocity relations in chip formation, 25

width of cut, relation to tube wall thickness (depth of cut) and side cutting edge angle, 142
work material flow stress properties,
  cold working, allowing for effects of, 203
  comparison of results of obtained from machining and dynamic shear tests, 89–90
  comparison of results of obtained from machining and high and low speed tension tests, 59–61, 88–89
  determination of from high speed compression test results, 94–96, 223–224
  determination of from machining test results, 56–59, 74–78, 91–93
  effect of dynamic strain ageing on, 93, 96
  effect of strain-rate on, 56–61, 77–78
  effect of strain-rate, temperature and velocity modified temperature on, 87–96
  equation of state, assumption in relation to, 75
  most appropriate methods for measuring and representation of, 206
  relating of, for plane strain and uniaxial conditions, 222
  representation of, for a range of plain carbon steels as a function of velocity modified temperature and carbon content, 223–227
work material thermal properties, 86–87